91446

D1327836
MANAGEMENT

Tools of Total Quality

OTHER STATISTICS TEXTS FROM
CHAPMAN AND HALL

Practical Statistics for Medical Research
Douglas Altman
The Analysis of Time Series – An Introduction
C. Chatfield
Problem Solving: A Statistician's Guide
C. Chatfield
Statistics for Technology
C. Chatfield
Introduction to Multivariate Analysis
C. Chatfield and A. J. Collins
Applied Statistics
D. R. Cox and E. J. Snell
An Introduction to Statistical Modelling
A. J. Dobson
Introduction to Optimization Methods and their Application in Statistics
B. S. Everitt
Multivariate Statistics – A Practical Approach
B. Flury and H. Riedwyl
Readings in Decision Analysis
S. French
Multivariate Analysis of Variance and Repeated Measures
D. J. Hand and C. C. Taylor
Multivariate Statistical Methods – a primer
Bryan F. Manley
Statistical Methods in Agriculture and Experimental Biology
R. Mead and R. N. Curnow
Elements of Simulation
B. J. T. Morgan
Probability: Methods and Measurement
Anthony O'Hagan
Essential Statistics
D. G. Rees
Foundations of Statistics
D. G. Rees
Decision Analysis: A Bayesian Approach
J. Q. Smith
Applied Statistics: A Handbook of BMDP Analyses
E. J. Snell
Elementary Applications of Probability Theory
H. C. Tuckwell
Intermediate Statistical Methods
G. B. Wetherill

Further information of the complete range of Chapman and Hall *statistics books is available from the publishers.*

Tools of Total Quality

An introduction to statistical process control

P. Lyonnet

Lecturer in Mechanical Engineering
University Institute of Technology
St. Denis
France

English translation by
Jack Howlett

CHAPMAN & HALL
London · New York · Tokyo · Melbourne · Madras

UK Chapman & Hall, 2–6 Boundary Row, London SE1 8HN

USA Van Nostrand Reinhold, 115 5th Avenue, New York NY10003

JAPAN Chapman & Hall Japan, Thomson Publishing Japan, Hirakawacho Nemoto Building, 7F, 1-7-11 Hirakawa-cho, Chiyoda-ku, Tokyo 102

AUSTRALIA Chapman & Hall Australia, Thomas Nelson Australia, 102 Dodds Street, South Melbourne, Victoria 3205

INDIA Chapman & Hall India, R. Seshadri, 32 Second Main Road, CIT East, Madras 600 035

English edition 1991

Typeset in 10/12 Times by KEYTEC, Bridport, Dorset
Printed and bound in Great Britain by
T J Press (Padstow) Ltd, Padstow, Cornwall

ISBN 0 412 37690 3 0 442 31248 2 (USA)

British Library Cataloguing in Publication Data
Lyonnet, P. (Patrick)
Tools of total quality.
1. Engineering industries. Quality control
I. Title II[Outils de la qualite total. English]
620.0045

ISBN 0-412-37690-3

Library of Congress Cataloging-in-Publication Data
Lyonnet, P. (Patrick)
[Outils de la qualité totale. English]
Tools of total quality : an introduction to statistical process control / P. Lyonnet ; English translation by Jack Howlett.
p. cm.
Translation of: Les outils de la qualité totale.
Includes bibliographical references.
Includes index.
ISBN 0-442-31248-2
1. Process control–Statistical methods. 2. Quality control–Statistical methods. I. Title.
TS156.8.L9613 1991
658.5′62′015195–dc20 90-46726
CIP

Contents

Preface **vii**

1 General questions and concepts **1**
1.1 Introduction 1
1.2 Why quality? 1
1.3 What is total quality? 5

2 Reliability in the choice of technology **10**
2.1 Quantitative analysis 10
2.2 Qualitative analysis 45

3 Controlling the manufacturing process **56**
3.1 Variability in manufactured products 56
3.2 Monitoring the manufacture 61
3.3 Interval between control actions 77

4 Quality control of goods received **82**
4.1 Control by attributes 82
4.2 Sequential testing 95
4.3 Control by measured properties 97
4.4 Sampling procedures 97

5 Cause-and-effect analysis **99**
5.1 The Ishikawa cause–effect diagram 99
5.2 Pareto or ABC analysis 101
5.3 Rank correlation: Spearman's coefficient 104
5.4 Analysis of variance 106
5.5 Experimental designs of type 2^n 112

6 Basic mathematics **118**
6.1 Probability: theory, definitions 118
6.2 Probability laws 120
6.3 Confidence interval for the mean 130
6.4 Linear regression 132

Exercises **136**

Solutions **148**

Appendices (Tables) **154**
1. Gaussian (normal) distribution 154
2. Student t distribution 157
3. χ^2 distribution 160
4. The F (Fisher–Snedecor) distribution 163
5. MTBF for a system following the Weibull law 168
6. Median ranks (Johnson's table) 170
7. Laplace transforms 178
8. Random numbers 179
9. Gamma law 181

Index **183**

Preface

For a long time, quality has been one of industry's main preoccupations. It remains so today.

There is some foundation for the statement that there is a 'quality crisis' in Europe, the methods traditionally used in our industries being unable to meet today's demands. Consequently it is essential to look for new directions in which to progress, taking account of the methods for achieving quality that have been developed in recent years. These methods impact on all parts of the industrial enterprise – marketing, manufacturing, research and development, after-sales services. All staff, administrative or technical, are involved.

The present book describes the tools that can help anyone who is concerned with the concept of 'total quality'; it will also be a valuable educational aid for students reading for degrees or other qualifications in engineering.

P. Lyonnet

1

General questions and concepts

1 INTRODUCTION: HOW CAN WE ACHIEVE TOTAL QUALITY?

This book is concerned with the various techniques and methods of analysis that can be used to ensure total quality in a project. In this first chapter we show the costs that result from not achieving quality, so as to make clear how important a quality-assurance service is to any enterprise. We stress also the involvement of marketing, particularly in laying down specifications for reliability.

Current techniques concerning reliability are developed in Chapter 2; here, much space is given to quantitative analysis, for this enables the whole range of problems that can be raised by questions of reliability to be dealt with in a co-ordinated manner. The advantage of using reliability as a guide to the choice of a technology is stressed.

Chapter 3 is devoted to methods for controlling manufacturing processes, in particular the implementation of control charts.

Chapter 4, effectively a continuation of Chapter 3, deals with the management of input of raw materials and other supplies and output of finished products. All the methods proposed are described in sufficient detail to make immediate application possible.

Chapter 5 develops methods for causal analysis – inferring causes from observed effects. The role of these methods in various parts of the enterprise is discussed, and their use in 'quality circles'.

Chapter 6 gives the mathematical apparatus needed to support the techniques described in the previous chapters, and the book concludes with a set of exercises, followed by their solutions; we trust that this will add to its educational value.

1.2 WHY QUALITY?

For a long time, quality specialists were accused of being more concerned with formalities than with productive work; this is no longer the case, but nevertheless it is important to emphasize areas where quality can prove a source of profit.

1.2.1 Measures of 'non-quality'

(a) *Customer action*
If a product fails to meet its specification during the guarantee period, costs are incurred that are easily measurable:

- involvement of the after-sales service;
- costs of repair or modification;
- transport costs;
- time wasted;
- possibility of penalty payments to the customer.

The cost of not achieving quality can be determined from these.

(b) *Losses within the enterprise*
Losses within the enterprise are mainly the costs of scrapping or reworking the product; they are easily determined from the costs of materials and labour.

Example
A sheet metal workshop produces laminations for rotors and stators of electric motors, with a value of 1 F per kilogram. A monthly loss of 2 tonnes, because of poor quality, would cost £3000, which would justify employing several staff purely for quality control.

(c) *Quality level imposed by the customer*
There are some markets in which a supplier is not allowed to compete unless he has a quality-control system that ensures that a stated level is reached. Such markets are aerospace, defence, nuclear reactors, national electric power stations etc.; they can insist on

- certification by the appropriate standards body;
- quality-assurance manuals, drawn up by a specialist organization;
- manufacturing control documentation, to show that the required standards have been adhered to;
- development of a checking procedure.

(d) *Influence of quality on sales; index of quality*
Obviously the quality of a product will affect the demand for it. A customer who has been let down by the performance of a product is not a good advertisement for the manufacturer, and sales are likely to suffer

in consequence. Whilst it may not be easy to quantify this effect it is nevertheless of real importance, and some indicators have been devised to help justify the cost of a quality-assurance service.

A conventional scale for denoting quality is shown in Fig. 1.1. The aim is to achieve an overall level of around 1 for a product. The best procedure is to estimate values of this index for the important features such as reliability, maintainability, aesthetics etc., with weighting factors indicating their relative importance, and to calculate an overall index from these; this will reveal the main weaknesses. Table 1.1 is an example.

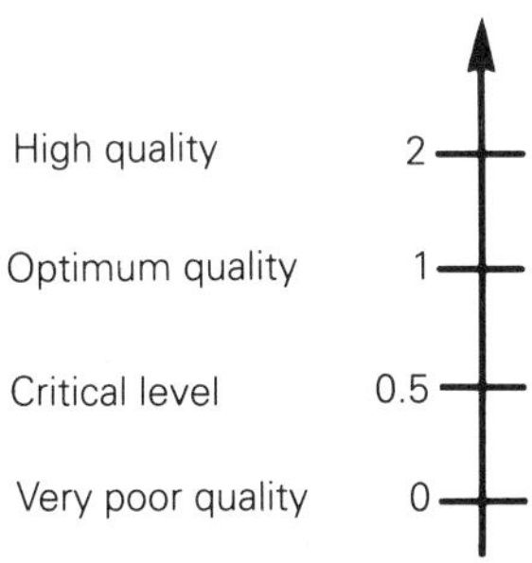

Figure 1.1 Conventional scale for quality.

Table 1.1 Weightings for separate features

Feature	Estimated index	Weight (%)	Overall index
Reliability	1.1	100	
Maintainability	0.8	80	
Performance	1.2	100	
Support costs	0.9	90	
Aesthetics	1.5	60	1.08

(e) *Quality–price–demand relations*

As we have noted, good quality will increase the demand for a product, poor quality will depress this; but demand will be depressed also by a high price. If a product sells badly because of poor quality the price will have to be lowered in order to increase sales. The situation can be represented by a set of curves as in Fig. 1.2.

A study of these relations between quality, price and demand enables the importance of the quality of a product to be assessed. It is important

Figure 1.2 Quality versus price and demand.

to recognize that for some products quality is the overriding consideration whilst for others the overriding consideration is price; thus one speaks of demand being *inelastic with respect to quality* or *inelastic with respect to price*.

1.2.2 Evolution of the quality concept

In the industrial world the concept of quality is a recent development; up to the end of the Second World War, in fact, it was scarcely taken into consideration. The various stages in its evolution often mirror the varying level of its adoption in industrial enterprises. We can distinguish the following chronological stages.

(a) *Production*
There was no quality service; everything was dominated by the manufacturing process. To ensure that the products delivered conformed to their specifications a service independent of the production organization weeded out those that did not.

(b) *Statistical control*
Statistical control became common during the 1950s, particularly with the appearance of the Military Standards tables.

(c) *Quality-assurance process*
The quality-assurance process was introduced into manufacturing, aligning production machinery with the requirements of product specifications. 'Control charts' and, particularly in Japan, 'quality circles' were introduced.

(d) *The quality-assurance concept*
Recognition of the possibilities opened up by quality assurance led to a reconsideration of the product in terms of feasibility of achievement; and this in turn led to decisions based on both quality and feasibility. The term 'customer satisfaction' came into use.

(e) *The 'total quality' concept*
Quality is now an important consideration both 'upstream' (marketing, production) and 'downstream' (sales, after-sales service): so the loop is closed. The concept first appeared in the USA with Feigenbaum (total quality control); this was an important breakthrough, all parts of the enterprise now being involved.

(f) *Achievement of total quality: 'total quality-control system'*
A total quality-control system involves bringing into play all the techniques that can affect the quality of the product.

'Meeting the customer's needs' means understanding his problems: there is often a world of difference between the stated need and the real need. It is marketing's business to resolve this difficulty and draw up an appropriate specification. This is then studied by the research and development organization who decide on the technical methods to be employed and make a provisional forecast of the reliability of the product; further control of quality is exercised in the manufacturing process, and final control is at the stage of product release. The quality-assurance task is completed in the after-sales and maintenance services.

It must never be forgotten that any enterprise is, above all, a collection of people: production will be affected by their health and quality of life; paying attention to these is a part of the quality process.

1.3 WHAT IS TOTAL QUALITY?

1.3.1 Definition

'Meeting the needs of, or providing the service required by, the customer or the user.'

Hidden in this statement are a number of points to which serious attention should be paid:

- the reliability of the product or service;
- the performance characteristics (of the product);
- its durability;
- its maintainability;

- its security;
- its effect on its environment (which must be acceptable);
- the ownership costs.

These correspond to the AFNOR (the French official standards body) definition of quality; to extend the definition to total quality we must add consideration of the extent to which production of the product or provision of the service contributes to the satisfaction of the people involved in the enterprise: the shareholders and the staff. Thus, all parts of the enterprise are involved (Figs 1.3 and 1.4).

1.3.2 Terms used in connection with quality

(a) *Quality assurance*
The accepted meaning of quality assurance is the laying down of a consistent set of standards and actions aimed at giving confidence in the achievement of quality. In practice this means compiling a quality manual and making sure that the engineer responsible for quality follows the key requirements in it.

(b) *Quality audit*
A quality audit is the detailed examination of

- the product,
- the manufacturing process and
- the organization, in the context of quality.

The existence of a quality standard is implied here.

(c) *Certification*
Certification involves the formal declaration by a recognized body that a product, a service or an enterprise meets a stated level of quality. It may take the form of the issue of a certificate or the authorization for the product to carry a particular label.

(d) *Marketing*
The role of marketing is crucial: for one thing it is upstream of manufacture, investigating where and in what volume the product will be sold, and for another it draws up the specification. A badly defined specification can involve the supplier in serious costs resulting from customer dissatisfaction and correction of errors.

It is also the business of marketing to ensure that the customers are aware of, and appreciate, the quality of the product.

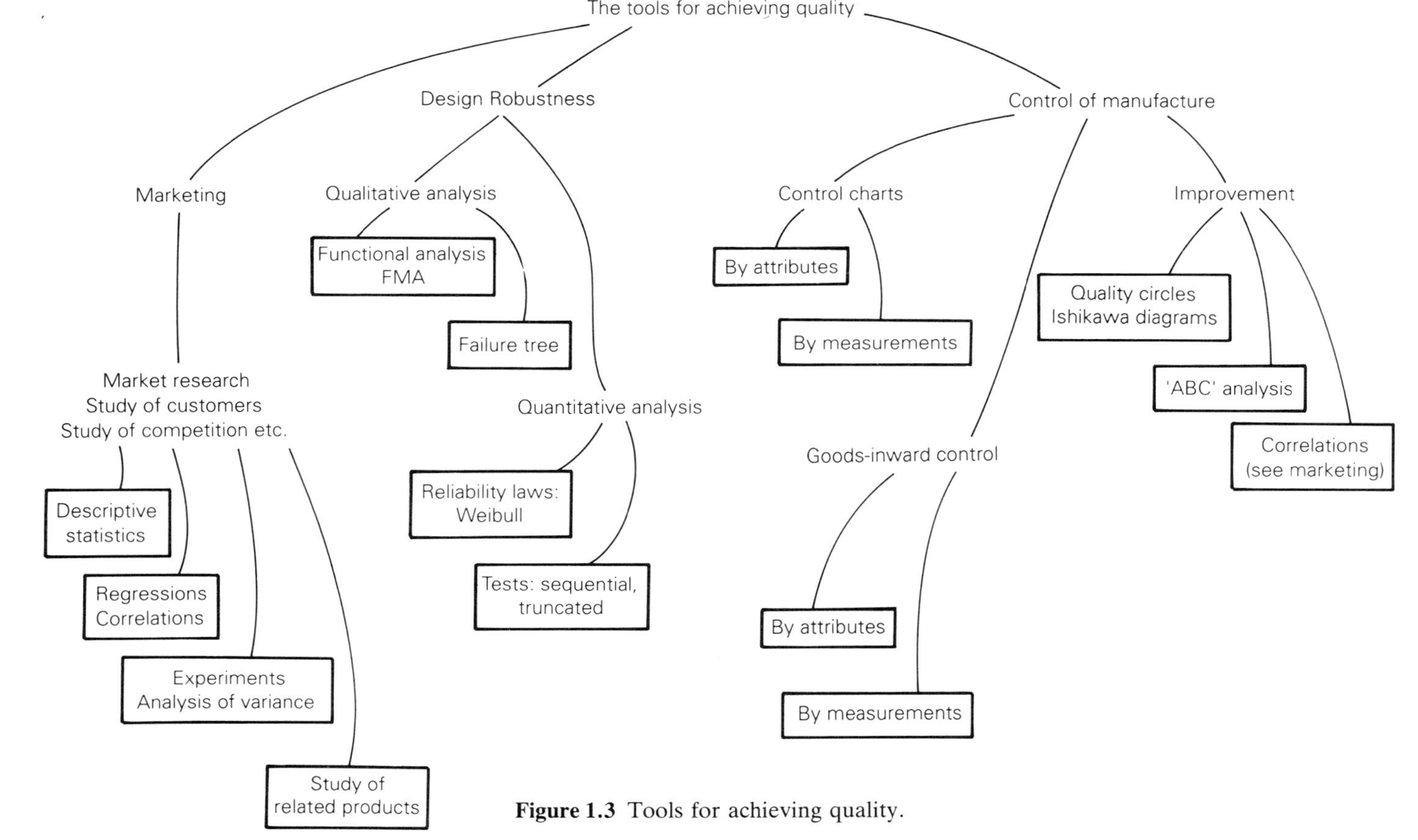

Figure 1.3 Tools for achieving quality.

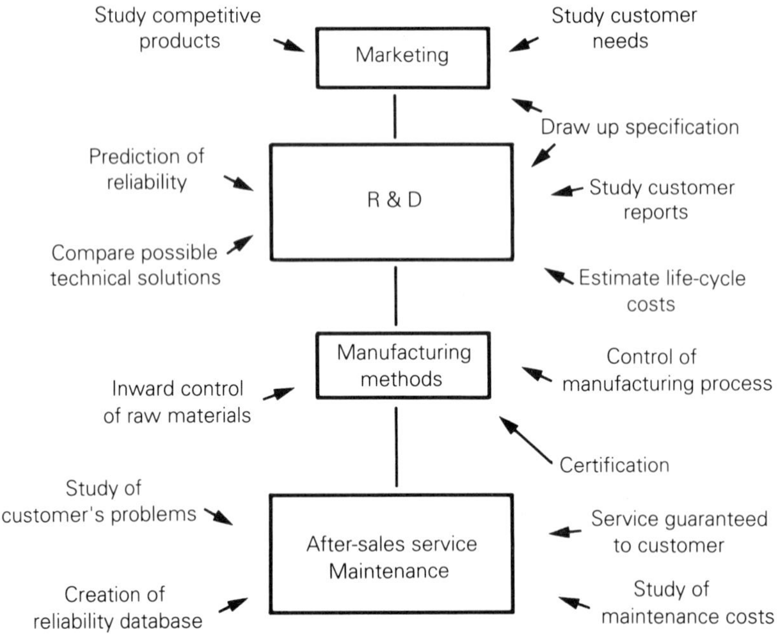

Figure 1.4 Activities required for achieving total quality.

(e) *Quality diagnosis*
Quality diagnosis is a matter of ensuring awareness of the key points concerning quality, which are as follows.

Costs resulting from poor performance These are distributed among the various parts of the enterprise:

- internal – scrap, reworking, injuries, removal of pollutants etc.
- external – costs of meeting guarantees and of providing after-sales services, penalties for late delivery etc.
- detection of faults and out-of-course events; post-delivery checks and tests

The overall cost of lack of quality is the sum of these separate costs.

Technical backwardness This is the failure of the enterprise to keep abreast of technical developments – for example, numerical control, use of information technology. It can result in loss of market share.

Level of customer satisfaction A questionnaire can help to give an idea of the enterprise's image in the marketplace, and of the return from the costs incurred in providing reliability, maintenance, performance etc.

Overall quality can be expressed as the product of its components:

$$Q(\text{overall}) = Q(\text{specification}) \times Q(\text{design}) \times Q(\text{execution}) \times Q(\text{exploitation})$$

(f) *Ownership costs*
It is not sufficient, when considering the purchase of a machine or a product, simply to compare the cost-to-performance ratios of the various possibilities: maintenance costs must also be taken into account. This explains the importance of predictions of reliability as an aid to making the choice.

(g) *Quality circles, quality tools*
A quality circle in an enterprise is a voluntary group of six to eight staff, with a leader, who meet regularly to discuss and if possible to solve problems, technical and other, concerning quality. They will always have real-life situations in mind, while aiming at the ideal.

Such a group will often be concerned with cause-and-effect investigations and with priorities.

2

Reliability in the choice of technology

The first essential is that the specification is fully defined, for on this the success of the project depends. This done, the reliability or 'operational security' of each possible technology must be investigated, so as to make possible the optimum choice. The term operational security (OS), which is coming more and more into use, means more than reliability and is understood as including

- *reliability* – ability to work without failure
- *maintainability* – ability to be restored quickly to working condition after failure
- *availability* – being in working condition when required
- *security* – remaining safe in case of failure

2.1 QUANTITATIVE ANALYSIS: FAILURE RATES, RELIABILITY LAWS

The instantaneous failure rate $\lambda(t)$ is defined by the statement that the probability that the device under consideration will fail in the (infinitesimal) interval $(t, t + \mathrm{d}t)$, having operated without failure up to time t, is $\lambda(t)\,\mathrm{d}t$. The cumulative failure function $F(t)$ is the probability that the device has failed at least once before time t is reached, and the reliability function $R(t)$ is the probability that it has not failed up to this time, i.e. that it has operated reliably. Clearly

$$R(t) = 1 - F(t)$$

$F(t + \mathrm{d}t)$ is the probability that the device fails at least once up to time $t + \mathrm{d}t$, and this will happen either if it fails not later than time t or if it does not fail up to that time and fails in the interval $(t, t + \mathrm{d}t)$. This gives the relation

$$F(t + \mathrm{d}t) = F(t) + R(t)\,\lambda(t)\,\mathrm{d}t$$

i.e. since $R(t) = 1 - F(t)$,

$$\lambda(t)\mathrm{d}t = \frac{F(t + \mathrm{d}t) - F(t)}{1 - F(t)}$$

$$= \frac{\mathrm{d}F(t)}{1 - F(t)}$$

If we reckon time from $t = 0$, $F(t) = 0$ at $t = 0$, and so integrating from 0 to t we have

$$\int_0^t \lambda(\tau)\,\mathrm{d}\tau = -\ln[1 - F(t)] = -\ln R(t)$$

and hence

$$R(t) = \exp\left[-\int_0^t \lambda(\tau)\,\mathrm{d}\tau\right]$$

$$F(t) = 1 - \exp\left[-\int_0^t \lambda(\tau)\,\mathrm{d}\tau\right]$$

A third function, the failure probability density function $f(t)$, is

$$f(t) = \frac{\mathrm{d}F(t)}{\mathrm{d}t} = \lambda(t)\exp\left[-\int_0^t \lambda(\tau)\,\mathrm{d}\tau\right]$$

$$= R(t)\lambda(t)$$

The equations between $R(t)$, $F(t)$, $f(t)$ and $\lambda(t)$ are the most general expressions for the laws of reliability.

An important quantity is the average time of fault-free operation, or the mean time between failures, MTBF. This is the mathematical expectation of the time to fail:

$$\text{MTBF} = \int_0^\infty \tau f(\tau)\,\mathrm{d}\tau$$

If we integrate this by parts we get the equivalent expression

$$\text{MTBF} = \int_0^\infty R(\tau)\,\mathrm{d}\tau$$

2.1.1 Reliability models

(a) *Constant failure rate: the exponential law*

In general, electronic components that have reached a state of maturity show a constant failure rate; this is expressed by putting $\lambda(t)$ equal to a constant, say λ, so that

$$\int_0^t \lambda(\tau)\,\mathrm{d}\tau = \lambda t \qquad R(t) = \exp(-\lambda t)$$

Example 1

If $\lambda = 2 \times 10^{-6}$ failures per hour and $t = 500$ h, then $\lambda t = 0.001$. So $R(t = 500) = \exp(-0.001) = 0.999$ and MTBF $= \int_0^\infty \exp(-\lambda t)\,\mathrm{d}t = 1/\lambda = 5 \times 10^5$ h.

(b) *The log-normal model*

The log-normal model gives a good representation of mechanical fatigue or wear. The probability density function is

$$f(t) = \frac{1}{\sigma(2\pi)^{1/2}} \exp\left[-\frac{1}{2}\left(\frac{\ln t - m}{\sigma}\right)^2\right]\frac{1}{t}$$

where m and σ are the mean and standard deviation respectively of $\ln t$.

Calculations with this model are carried out most easily in terms of the reduced variable $u = (\ln t - m)/\sigma$, which is distributed normally with mean zero and standard deviation unity. Making this substitution we find

$$\text{MTBF} = \int_0^\infty t f(t)\,\mathrm{d}t = \exp(m + \tfrac{1}{2}\sigma^2)$$

Example 2

The lifetime of the con-rods of an automobile engine follows a log-normal law with parameters $m = 5$, $\sigma = 1.4$, time being measured in hours. Find (1) the reliability after 300 h and (2) the MTBF.

(1) $u = (\ln 300 - 5)/1.4 = 0.502$. From Table 1 (p. 154) we find that $F(u) = 0.692$, and so $R(t = 300) = 1 - F(0.502) = 0.308$ (which is poor).
(2) $m + \frac{1}{2}\sigma^2 = 5.98$, and so MTBF $= \exp(5.98) = 395$ h.

(c) *A more general law: the Weibull model*

(i) *General form of Weibull's law*

The most general form of Weibull's law includes many of the simpler models and is given by

$$R(t) = 1 - F(t) = \exp\left[-\left(\frac{t - \gamma}{\eta}\right)^\beta\right]$$

The instantaneous failure rate $\lambda(t)$ is

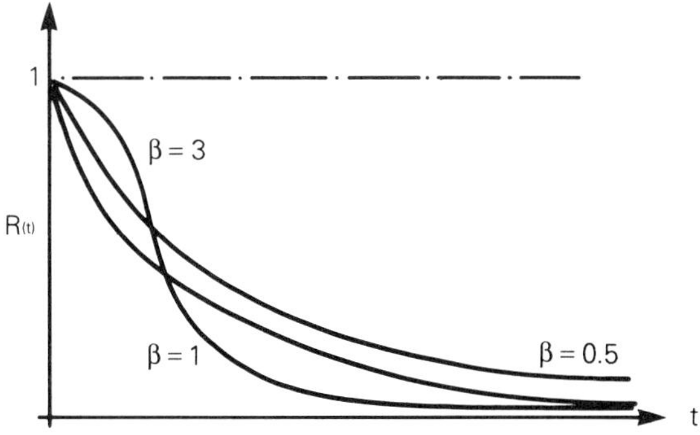

Figure 2.1 Examples of Weibull law.

$$\lambda(t) = \frac{f(t)}{R(t)} = \frac{\beta}{\eta}\left(\frac{t-\gamma}{\eta}\right)^{\beta-1}$$

β is the *shape* parameter, η is the *scale* parameter and γ is the *location* parameter. A study of the form of $\lambda(t)$ shows that

for $\beta < 1$ $\lambda(t)$ is a decreasing function of t
for $\beta = 1$ $\lambda(t)$ is constant, equal to $1/\eta$
for $\beta > 1$ $\lambda(t)$ is an increasing function of t

For the particular case $\beta = 1$, $\gamma = 1$, Weibull's law reduces to the exponential law:

$$R(t) = \exp(-\,t/\eta)$$

i.e. the exponential law with parameter $\lambda = 1/\eta$.

For $\beta \geqslant 3$ Weibull's law approximates to the normal law, more closely as β increases.

(ii) *Estimation of the parameters*

A basic problem in connection with the use of Weibull's law is estimating the values of the parameters β, η and γ for given data. Two methods are available:

- a purely numerical method, which leads to differential equations that are difficult to solve and consequently is little used;
- a graphical method which uses special paper, called Weibull paper, ruled with functional scales. This is the method most used, and the one we now describe.

The scales are as follows: ordinate, $Y = \ln\ln\{1/[1 - F(t)]\}$; abscissa, $X = \ln t$.

The case $\gamma = 0$ corresponds to the assumption that the origin of time is known and is given by the data. Then

$$R(t) = 1 - F(t) = \exp\left[-\left(\frac{t}{\eta}\right)^{\beta}\right]$$

and so

$$\ln\left[\frac{1}{1 - F(t)}\right] = \left(\frac{t}{\eta}\right)^{\beta}$$

$$Y = \ln\ln\left[\frac{1}{1 - F(t)}\right] = \beta \ln t - \beta \ln \eta$$

$$X = \ln t$$

X and Y are both functions of t; but if we put $A = \beta$, $B = \beta \ln \eta$, which are constants, we have

$$Y = AX - B$$

which is the equation of a straight line. Thus if a set of observations of $F(t)$ at a number of values of t are described by a Weibull law with $\gamma = 0$, then these should lie on a straight line on Weibull paper.

Figure 2.2 illustrates this. On this paper the origin for Y is the ordinate $F(t) = 0.632$ (or 63.2%), because if $Y = 0$ then $\ln\{1/[1 - F(t)]\} = 1$ and so $F(t) = 1 - 1/e = 0.632$.

The shape parameter β is the slope of the line. To find its value we draw a line through the point ($t = 1$, $F(t) = 0.632$) (marked on the paper) parallel to the line on which the data points lie and read the value at the intersection of this with the β scale.

The value of the scale parameter η is read at the intersection of the data line with the parallel to the X axis through the ordinate $F(t) = 0.632$, because there $Y = 0$ and so $AX - B = 0$, which from the definitions of A and B gives $X = \ln \eta$. Since $X = \ln t$, this point is $t = \eta$.

In the example of Fig. 2.2, $\beta = 1.5$ and $\eta = 20\,000$ h.

For the case $\gamma = 0$ which we are considering

$$\text{MTBF} = \int_0^{\infty} R(t)\,dt = \eta\Gamma(1 + 1/\beta)$$

where Γ denotes the gamma function: formulae and a table are given in Appendix 9. For the above example we find, for $\beta = 1.5$, $\Gamma(1 + 1/\beta) = 0.9027$, and so (to a realistic accuracy)

$$\text{MTBF} = 20\,000 \times 0.9027 = 18\,000 \text{ h}$$

For $\gamma > 0$ the data cannot be linearized by the above process; instead,

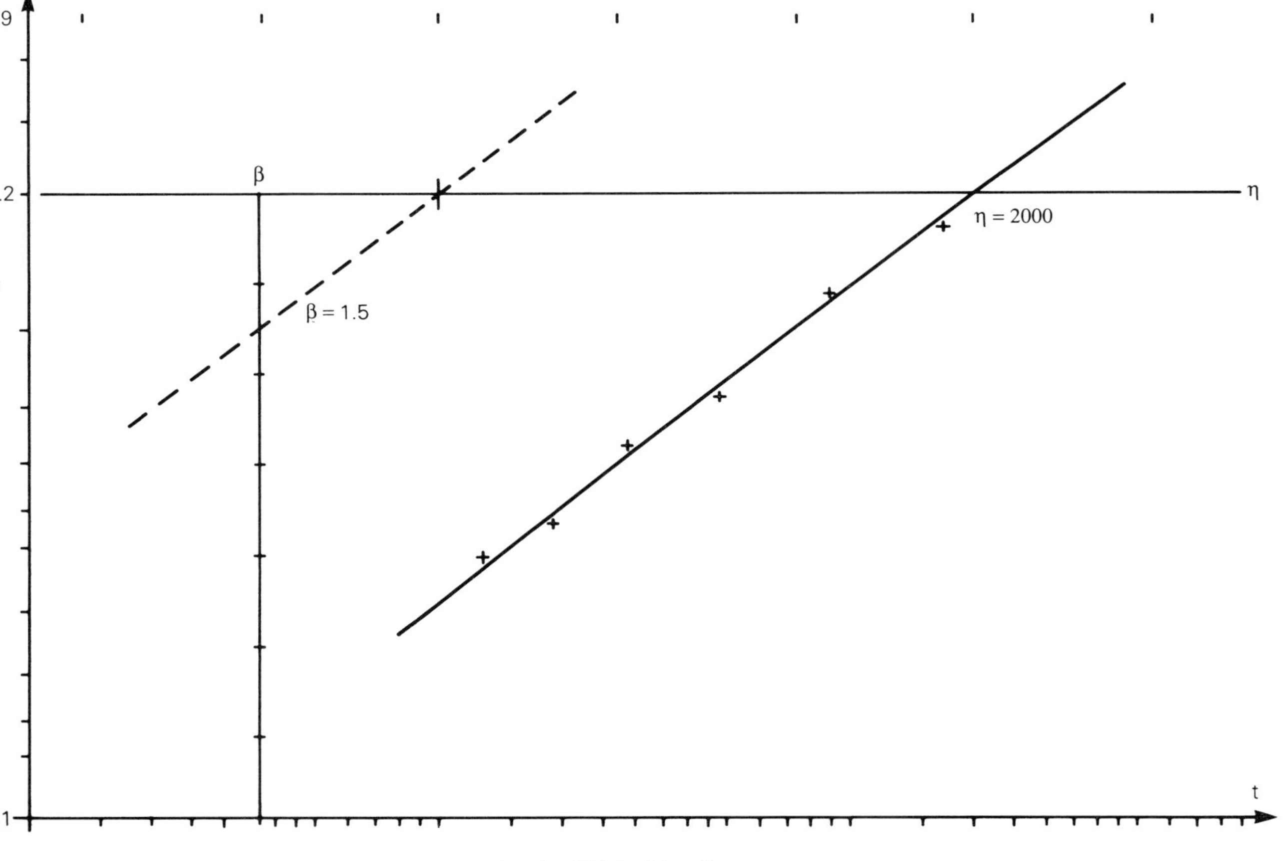

Figure 2.2 Fitting Weibull law ($\gamma = 0$).

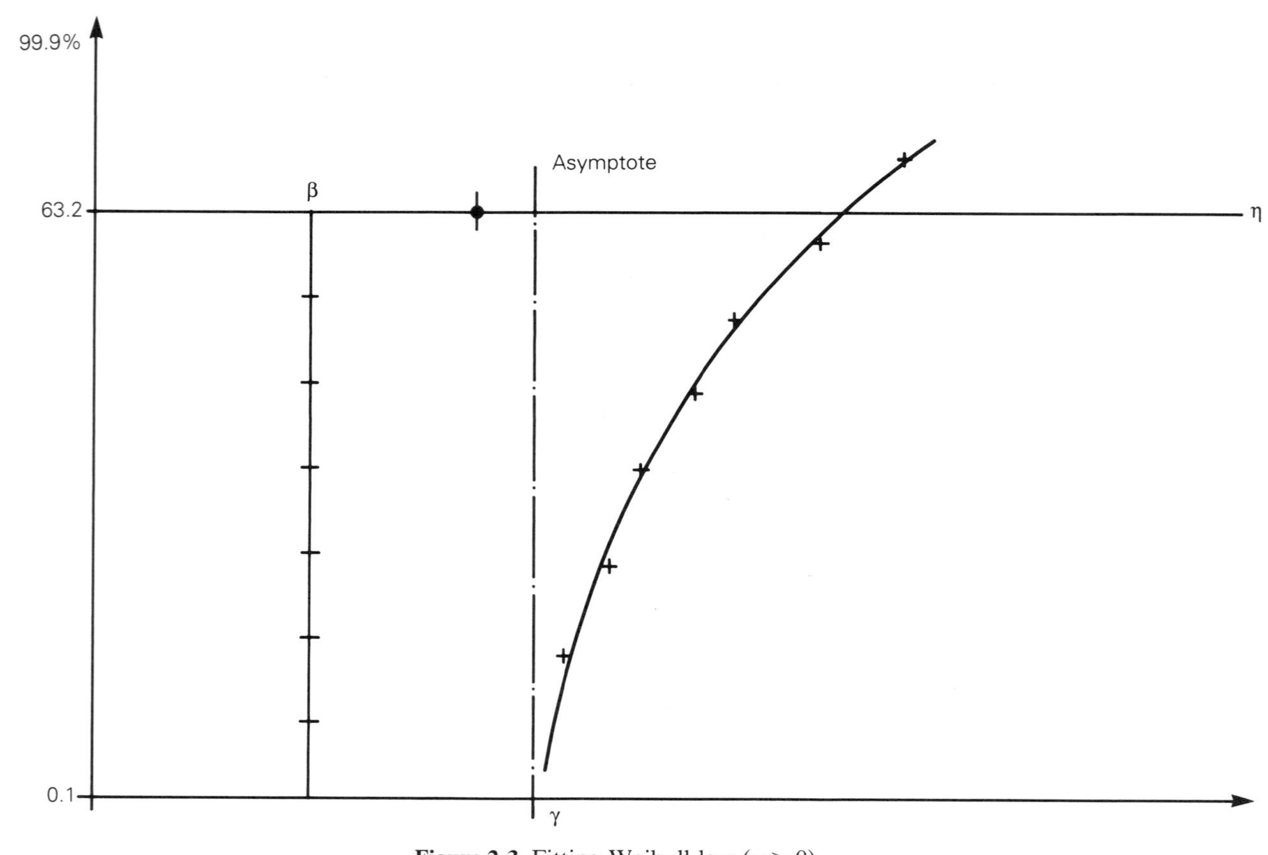

Figure 2.3 Fitting Weibull law ($\gamma > 0$).

the points will lie on a curve that has a vertical asymptote, and γ is given by the value of t at which this asymptote intersects the t axis. This follows from the fact that, when $t = \gamma$, $F(t) = 0$, and so $Y = \ln\ln 1 = \ln 0 = -\infty$. Figure 2.3 illustrates this.

To find the parameters in this case we first estimate the position of the asymptote, as in Fig. 2.3, and obtain a first estimate for γ, say γ'. We now change the t scale to $t' = t - \gamma'$ and repeat the linearization process with the new time scale. If this gives an acceptable approximation to a straight line then γ' is a sufficiently good approximation to the true value of γ and we can continue as before, finding β and η. If not, we go through the estimation process again, estimate a value γ'' and plot the data again with the time scale $t'' = t' - \gamma' = t - \gamma' - \gamma''$; and so on. But if a third repeat of this process does not give an adequately linear plot we must conclude that the data do not follow a Weibull law; they may follow a mixture of Weibull laws with different parameters, or some quite different law.

For $\gamma < 0$ we have $t - \gamma > 0$ for $t > 0$ and $F(t) \to 1 - \exp[-(-\gamma/\eta)]$ as $t \to 0$. Thus γ $(= \ln\ln 1/[1 - F(t)]) \to \beta \ln(-\gamma/\eta)$ as X $(= \ln t) \to -\infty$, i.e. the curve of Y against X has a horizontal asymptote.

One way of proceeding in this case is to try a succession of estimates for γ until an acceptably linear plot is obtained and then to continue as before. Figure 2.4 illustrates this.

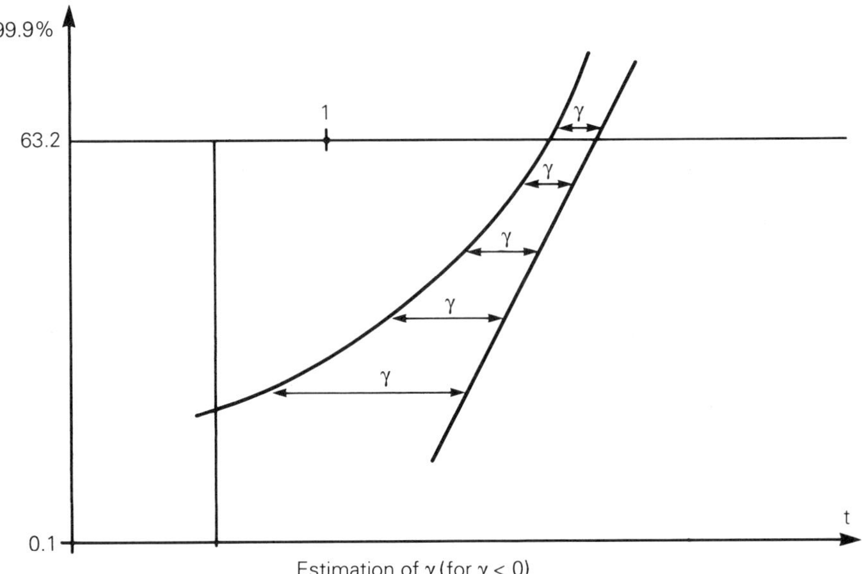

Figure 2.4 Fitting Weibull law ($\gamma < 0$).

(iii) *Another method for determining* γ

Using the same change of variables as before, an estimate for the value of γ can be computed as follows:

$$\gamma = X_m - \frac{(X_{max} - X_m)(X_m - X_{min})}{(X_{max} - X_m) - (X_m - X_{min})}$$

where X_{max} is the value of X corresponding to the maximum value Y_{max} of Y, X_{min} is the value of X corresponding to the minimum value Y_{min} of Y and X_m corresponds to the midpoint Y_m between Y_{max} and Y_{min} measured on a linear scale. This is illustrated in Fig. 2.5.

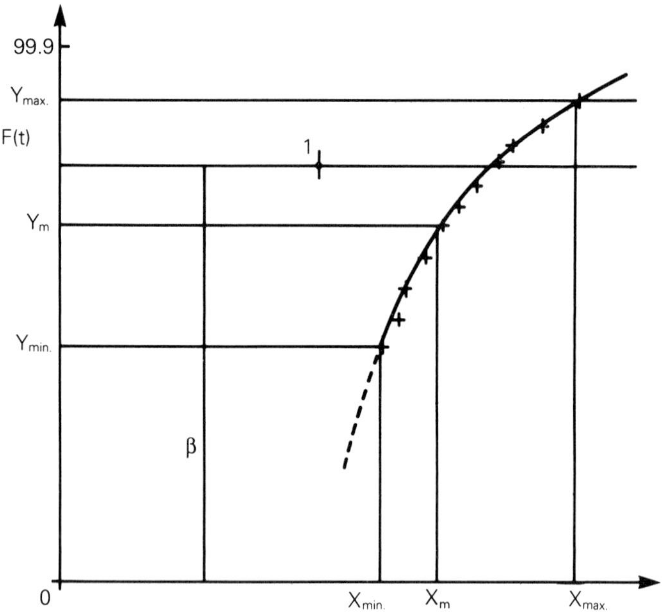

Figure 2.5 Weibull law: Estimation of γ using method in 2.1.1 (cxiii)

Example 3

Given that the following data can be represented by a general Weibull law, find the values of the three parameters Y, β and η.

lifetimes in hours:

705, 812, 902, 995, 1070, 1171, 1301, 1440, 1650

From these values

$$Y_{max} = 0.9 \Rightarrow X_{max} = 1650$$
$$Y_{min} = 0.1 \Rightarrow X_{min} = 705$$
$$Y_m = 0.38 \Rightarrow X_m = 980$$

Substituting in the above formula we obtain

$$\gamma = 980 - (1650 - 980)(980 - 705)/[(1650 - 980) - (980 - 705)]$$
$$= 513 \text{ h}$$

Knowing γ we can linearize the curve and find the other parameters; we obtain the values $\beta = 1.8$, $\eta = 700$ h.

Example 4
The lifetime values (h) of a mechanical system are as follows:

5, 112, 202, 295, 370, 471, 601, 740, 905

The corresponding pairs of values of Y and X are now (0.9, 905), (0.1, 5) and (0.38, 275), and the same calculation as in Example 3 gives

$$\gamma = 197 \text{ h} \qquad \beta = 1.8 \qquad \eta = 705 \text{ h}$$

(iv) *Estimation of F(t)*
There are two methods for estimating $F(t_i)$.

(1) Method of median ranks (for small samples):

$$F(i) = \frac{\Sigma n_i - 0.3}{n + 0.4}$$

There are tables of this (see pp. 170–177).

(2) Method of mean ranks (more commonly used):

$$F(t_i) = \frac{\Sigma n_i}{n + 1}$$

(v) *Application: an example*
A sample of nine ball bearings has been put into service as a test of a new production series. The results were the following lifetimes (h):

801, 312, 402, 205, 671, 1150, 940, 495, 570

(1) Assuming a Weibull law, find the parameters.
(2) Compute the MTBF.
(3) Find, both graphically and by computation, the reliability after 600 h.

(1) We start by putting the observations in increasing order and tabulating the distribution function

$$F(i) = \frac{i}{n + 1} = \frac{i}{10}$$

(Table 2.1). We next plot these on Weibull paper with lifetimes as abscissae and $F(i)$ as ordinates; this gives the graph in Fig. 2.6. From the figure we find $\gamma = 0$ (because the points lie on a straight line), $\beta = 1.8$ and $\eta = 710$ h.

(2) MTBF $= E(t) = \eta\Gamma(1 + 1/\beta) = 710 \times \Gamma(1.555) = 631$ h from Table 5.1.

(3) Computationally, $R(t = 600) = \exp[-(600/710)^{1.8}] = 0.480$. From the graph, at $t = 600$, $F(t) = 0.52$, and so $R(t) = 1 - F(t) = 0.48$.

Table 2.1 Observed data for Figure 2.6

Position	Lifetime	$F(i)\%$	Position	Lifetime	$F(i)\%$
1	205	10	6	671	60
2	312	20	7	801	70
3	402	30	8	940	80
4	495	40	9	1150	90
5	570	50			

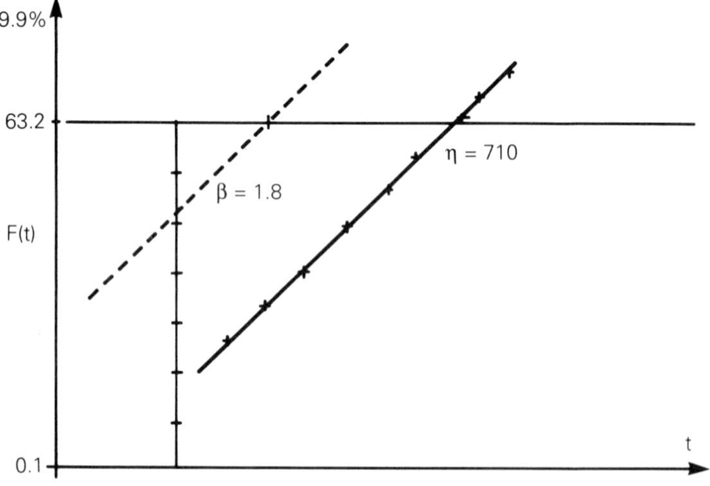

Figure 2.6 Weibull law: Fitting data from Table 2.1.

(d) *Mixture of Weibull laws (use of method of median ranks)*
A mixture of different Weibull laws may be needed to model a set of failure data because

- the items may have come from different populations
- several different modes of failure may coexist simultaneously

This is illustrated in Fig. 2.7 which shows three different populations coexisting.

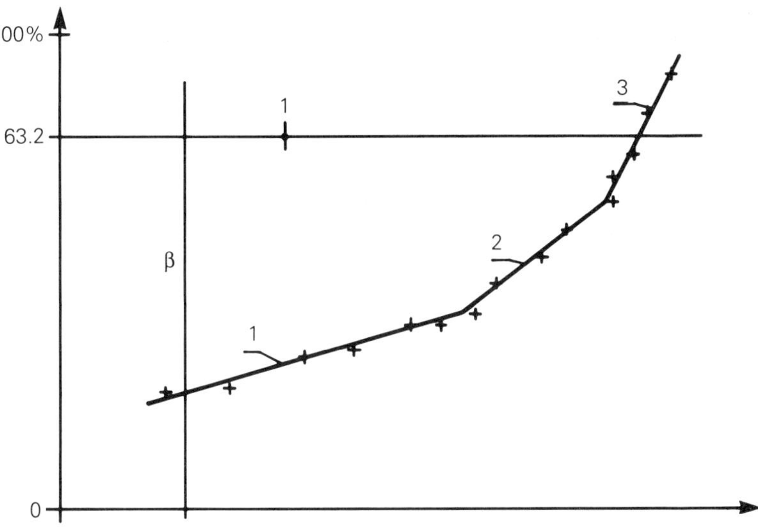

Figure 2.7 Weibull Law: Three populations coexisting.

(i) *Solution of the mixed law problem*
The data (times of fault-free operation) are plotted after applying the method of median ranks,

$$F(t_i) = \frac{\Sigma\, n_i - 0.3}{n + 0.4}$$

(or using Johnson's table, for small samples).

The size of each subpopulation is found by counting the number of items that belong to it, say n_1, n_2, n_3, and an estimate of the proportion of each in the total population is given by $P_1 = n_1/n$ etc. Each subpopulation is studied independently, either by plotting on Weibull paper or by computation; thus for the first we would have

$$F(t_{1i}) = \frac{\Sigma\, n_{1i} - 0.3}{n_1 + 0.4}$$

and similarly for the others.

The reliability model for this system with three subpopulations is then

$$R_{1,2,3}(t) = \sum P_i \exp\left[\left(\frac{t-\gamma_i}{\eta_i}\right)^{\beta_i}\right] \qquad i = 1, 2, 3$$

Example 5

The following lifetimes are observed, given in ascending order: 235, 390, 540, 580, 730, 766, 800, 850, 900, 940, 980, 1100, 1150, 1200, 1240, 1310, 1400, 1455. Find the model.

Plotting on Weibull paper indicates that there are two populations P_1 and P_2. For P_1 we have the values given in Table 2.2 and from the plot $\gamma = 0$, $\beta = 2.3$ and $\eta = 450$.

Table 2.2 Values for population P_1

Order	Lifetime	$F(t_i)$
1	235	20.5
2	390	50
3	540	79.41

For P_2 the values are given in Table 2.3, and $\gamma = 0$, $\beta = 4$ and $\eta = 1020$. Hence

$$R(t) = \frac{3}{19}\exp\left[-\left(\frac{t}{450}\right)^{2.3}\right] + \frac{16}{19}\exp\left[-\left(\frac{t}{1020}\right)^{4}\right]$$

Table 2.3 Values for population P_2

Order	Lifetime	$F(t_i)$	Order	Lifetime	$F(t_i)$
1	690	5.88	9	1020	52.94
2	730	11.76	10	1100	58.82
3	766	17.64	11	1150	64.70
4	800	23.53	12	1200	70.58
5	850	29.41	13	1240	76.47
6	900	35.29	14	1310	82.35
7	940	41.17	15	1400	88.23
8	980	47.06	16	1455	94.12

(e) *Program for finding the Weibull parameters, assuming that $\gamma = 0$*
The program uses the transformations

$$X_i = \ln t_i \qquad Y_i = \ln \ln \left[\frac{1}{1 - F(t_i)} \right]$$

Estimates of the parameters β and η (actually, η^β) are found by least-squares fitting of a straight line (Fig. 2.8).

```
10  CLS
20  DIM T(100)
30  PRINT "WEIBULL MODEL"
40  PRINT "ENTER TBF ONE AT A TIME"
50  PRINT "DO TBF=0 TO PROCESS DATA"
60  I=0
70  I=I+1
80  PRINT I
90  INPUT " TBF=";T(I)
100 IF T(I)=0 THEN 120
110 GOTO 70
120 N=I
130 I=0
140 T=0
150 F=0
160 T2=0
170 F2=0
180 F5=0
190 PRINT "number of TBF taken="
200 N=N-1
210 PRINT N
220 N=N+1
230 I=0
240 FOR I=1 TO N-1
250 T=T+LOG(T(I))
260 F1=1/(1-I/N)
270 F3=LOG(F1)
280 F=F+LOG(F3)
290 T2=T2+LOG(T(I))^2
300 F2=F2+LOG(F3)^2
310 F4=LOG(T(I))*LOG(F3)
320 F5=F5+F4
330 NEXT I
340 N=N-1
350 B=(F5-T*F/N)/(T2-T^2/N)
360 E=T/N-F/N/B
370 E1=EXP(E)
380 R1=F5-T*F/N
390 R2=(T2-T^2/N)*(F2-F^2/N)
```

Figure 2.8 BASIC program for computing Weibull parameters ($\gamma = 0$).

```
400  R3=SQR(R2)
410  R=R1/R3
420  CLS:PRINT “RESULT OF PROCESSING:”
430  PRINT“BETA=”;B
440  PRINT“ETA=”;E1
450  PRINT “LINEAR CORRELATION :R=”;R
460  PRINT “Compute MTBF if YES enter 0 if NO 20”
470  INPUT G
480  IF G=0 THEN GOTO 490 ELSE GOTO 570
490  X=0
500  P=.001
510  X=X+P
520  G=G+ABS(LOG(1/X))^(1/B)*P
530  IF X>=1 THEN 550
540  GOTO 510
550  X=E1*G
560  PRINT“MTBF=”;X
570  PRINT “Compute the R(T) and F(T)”
580  INPUT “to compute R(T) and F(T) enter t if NO do 0”; T3
590  IF T3=0 THEN GOTO 670
600  R1=(T3/E1)^B
610  R2=EXP(−R1)
620  PRINT “For t=”;T3;“R(T) is equal to”;R2
630  R4=1−R2
640  PRINT “For t=”;T3;“F(T) is equal to”;R4
650  INPUT “To repeat calculation enter t if NO 0”;T3
660  IF T3=0 THEN GOTO 670 ELSE GOTO 590
670  END
```

Figure 2.8 (continued)

2.1.2 Verification of the models

Any model constructed in a reliability study will be based on a sample drawn from the population being investigated, and some assumption will always be made concerning the distribution law for that population – exponential, log normal etc. There is therefore the question of the validity of this assumption, and this can be answered by applying what are called goodness-of-fit tests. In using these statistical tests we must always recognize that there is a risk of being wrong, measured by the probability α that the test will give the wrong result. α is called the significance level of the test, and we aim to make its value small.

(a) *The chi-squared (χ^2) test*

The condition for the χ^2 test to be applicable is that there are at least 50 observations: $n \geqslant 50$. It is usual to group the observations into classes so that there are at least five in each class; the classes need not be at regular intervals.

The test is based on the differences between the numbers of observations in each class and the number predicted by the model; the measure of this difference used by the test is

$$E = \sum_{1}^{r} \frac{(n_i - np_i)^2}{np_i}$$

where r is the number of classes, n_i is the number of observations in class i, n is the total number in the sample ($= \Sigma\, n_i$), p_i is the probability that an observation will be in class i and np_i is the expected (theoretical) number in class i. E is distributed approximately according to the χ^2 law with v degrees of freedom where $v = r - k - 1$ and k is the number of parameters whose values have to be estimated in deriving the model. This depends on the underlying law assumed: for example, $k = 1$ for the exponential law, $k = 2$ for the normal (Gaussian) law and $k = 3$ for the Weibull law.

χ^2 is a function of two variables, the degrees of freedom v and the significance level α; for given α

$$P(E > \chi^2_{v,1-\alpha}) = 1 - \alpha$$

and the test is that if $E > \chi^2_{v,1-\alpha}$ we reject the hypothesis on which our theoretical model is based.

It should be noted that other tests can be used, e.g. the Kolmogorov–Smirnov test (see Lyonnet, *Maintenance Planning*, Chapman & Hall).

2.1.3 Predicting reliability

In order to estimate the cost of a maintenance service and to decide how to implement this, it is important to be able to predict the reliability of the equipment that is to be maintained; this is also important for the choice of techniques to be used. Published data are available on which the prediction can be based.

(a) *Relevant databases*

There are tables relating to currently used electronic and mechanical components; those most commonly used are the following: in France, those published by CNET (NPRD 1 and 2) and by EDF; in the USA, Rome Air Development Center (RADC), NASA, US Navy (FARADA) and AVCO Corporation. The tables give

- the name (identifier) of the item,
- the MTBF,
- the failure rate, either average or calculated on the assumption of a constant $\lambda(t)$,
- basic statistical information (e.g. confidence intervals),
- a multiplier to be applied to the given failure rate when the equipment is used under each of a number of stated conditions.

Two comments are relevant.

1. There are fundamental differences between electronic and mechanical systems.

   For electronic components the statistical information is more important than for mechanical components and the MTBF derived from this is more reliable. The failure rates are usually constant and can be taken as the values given in the tables.

   This does not hold for mechanical components. In practice, failure rates are found not to be constant, mechanical components are less well differentiated than electronic components and there is less statistical information.
2. The conditions under which the equipment is actually used are often very different from those assumed in the tables. The multipliers that should be applied in the various environments do not always take into account

   - the installation conditions
   - vibration
   - temperature
   - dust
   - corrosion
   - mechanical constraints

The general conclusion is that results obtained on the basis of the published tables should be treated with caution, especially in the case of mechanical systems.

For illustration, Table 2.4 gives an extract from the CNET NPRD-2 and RADC-NPRD-3 tables.

(i) *The RADC NPRD-3 tables*

These provide a databank for mechanical systems and reliability figures are included. The headings have the following meanings:

CLASS	a family of components having the same function
TYPE	the particular member of that family

Table 2.4 Extract from CNET NPRD-2 and RADC NPRD-3 tables

Failure rate/10^6 h

Environment	Application		$\hat{\lambda}$	60% upper single-sided confidence	60% confidence interval		Number of records	Number failed	Operating hours (10^6)
	MIL	COML			Lower	Upper			

Part class : Compressor
Type : Air

Failure rate/10^6 h

Environment	Application		$\hat{\lambda}$	60% upper single-sided confidence	60% confidence interval		Number of records	Number failed	Operating hours ($\times 10^6$)
	MIL	COML			Lower	Upper			
GRM	x		5.959	–	4.793	7.424	1	19	3.188
SHS	x		720.694	–	633.177	821.659	1	49	0.067

Part class : Compressor
Type : General

Failure rate/10^6 h

Environment	Application		$\hat{\lambda}$	60% upper single-sided confidence	60% confidence interval		Number of records	Number failed	Operating hours ($\times 10^6$)
	MIL	COML			Lower	Upper			
DOR	x		–	3.742	–	–	1	0	0.244
AU	x		1992.793	–	1942.226	2044.922	1	1 106	0.555

Part class : Brake
Type : General

Failure rate/10^6 h

Environment	Application		$\hat{\lambda}$	60% upper single-sided confidence	60% confidence interval		Number of records	Number failed	Operating hours ($\times 10^6$)
	MIL	COML			Lower	Upper			
GRF	x		4.274	–	0.847	12.995	1	1	0.234
A	x		766.250	–	760.349	772.207	1	11,964	15.615
AU	x		213.143	–	209.249	217.123	1	2,131	9.998
AUT		x	11.570	–	7.835	16.976	3	7	0.605
HEL	x		100.00	–	94.333	106.062	1	223	2.230

Part class : Bearing
Type : Ball

Failure rate/10^6 h

Environment	Application		$\hat{\lambda}$	60% upper single-sided confidence	60% confidence interval		Number of records	Number failed	Operating hours ($\times 10^6$)
	MIL	COML			Lower	Upper			
DOR	x		0.010	—	0.007	0.014	3	9	903.040
SAT	x		—	0.688	—	—	2	0	1.332
GRF	x		1.148	—	1.001	1.319	8	44	38.320
GRF		x	13.975	—	10.356	19.410	1	9	0.644
GRM	x		0.094	—	0.054	0.159	1	4	42.554
A	x		5.133	—	4.787	5.507	2	158	30.784
A		x	1.372	—	0.272	4.171	1	1	0.729
AI	x		4.829	—	3.799	6.148	1	16	3.313
HEL	x		13.398	—	10.963	16.408	2	22	1.642
SHS	x		—	0.053	—	—	2	0	17.156
SUB	x		4.728	—	1.923	10.220	1	2	0.423

Part class : Bearing
Type : Bushing

Failure rate/10^6 h

Environment	Application		$\hat{\lambda}$	60% upper single-sided confidence	60% confidence interval		Number of records	Number failed	Operating hours ($\times 10^6$)
	MIL	COML			Lower	Upper			
GRF		x	–	0.046	–	–	7	0	19.922
A	x		–	0.609	–	–	1	0	1.503
A	x		–	1.020	–	–	1	0	0.898
HEL	x		21.146	–	20.148	22.202	2	321	15.180

Part class : Bellows
Type : Diaphragm burst

Failure rate/10^6 h

Environment	Application		$\hat{\lambda}$	60% upper single-sided confidence	60% confidence interval		Number of records	Number failed	Operating hours ($\times 10^6$)
	MIL	COML			Lower	Upper			
DOR	x		–	1.384	–	–	1	0	0.662

Part class : Bellows
Type : Explosive

Failure rate/10^6 h

Environment	Application		$\hat{\lambda}$	60% upper single-sided confidence	60% confidence interval		Number of records	Number failed	Operating hours ($\times 10^6$)
	MIL	COML			Lower	Upper			
DOR	x		—	0.014	—	—	1	-	65.600

Part class : Bellows
Type : General

Failure rate/10^6 h

Environment	Application		$\hat{\lambda}$	60% upper single-sided confidence	60% confidence interval		Number of records	Number failed	Operating hours ($\times 10^6$)
	MIL	COML			Lower	Upper			
DOR	x		—	0.068	—	—	1	0	13.520
GRF	x		—	65.429	—	—	1	0	0.014

ENVIRONMENT	coded as follows
DOR	Dormant: the item is connected to the system but is out of operation for long periods
SAT	Satellite: in orbit around the earth; no access for maintenance
GRF	Fixed terrestrial installation; permanent; ventilated; maintenance by military personnel
GRM	Mobile terrestrial installation; conditions harsher than for GRS – vibration, shocks; maintenance more difficult

(b) *System reliability*

From the point of view of reliability the aims in constructing a system made up of a number of components are as follows:

- to satisfy the customer's requirements as expressed in the reliability specification, or his needs if not so specified;
- to choose an appropriate technology, using the reliability cost ratio as criterion;
- to improve the reliability by bringing to light the critical points; it should be possible to make a prediction of the reliability in the design stage.

A system consists of a set of elements or subsystems each of which provides one or more stated functions; thus the design proceeds by breaking the system down into elements, for each of which a numerical value for the reliability can be given, and then constructing a representation of the organization of these together to form the complete system. This is called constructing a block diagram for the system.

(i) *Block-schematic reliability calculation*

A *series system* (Fig. 2.9) fails if any one of its components or subsystems fails; if $R_i(t)$ is the reliability function for component or subsystem i, and if all are independent, the system reliability $R_s(t)$ is

$$R_s(t) = R_1(t) \times R_2(t) \times \ldots \times R_n(t) = \Pi_i R_i(t)$$

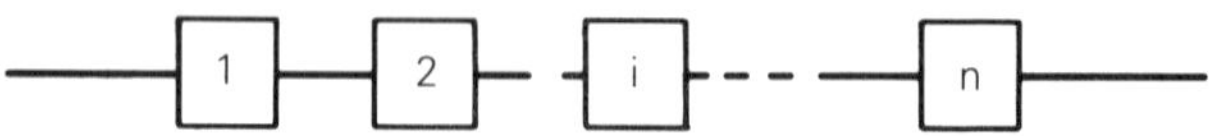

Figure 2.9 Block schematic for serial system.

A *parallel system* (Fig. 2.10) fails if and only if every one of its components or subsystems fails; since failure corresponds to the failure function $F(t) = 1$, the equation now is

$$F_s(t) = \Pi_i F_i(t)$$

or, since $R(t) = 1 - F(t)$,

$$R_s(t) = 1 - \Pi_i[1 - R_i(t)]$$

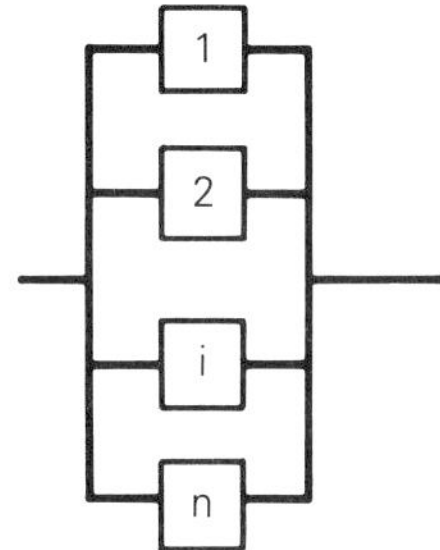

Figure 2.10 Block schematic for parallel system.

A general system can always be represented as a collection of series and parallel subsystems, themselves connected in series and/or parallel. Thus for the system in Fig. 2.11

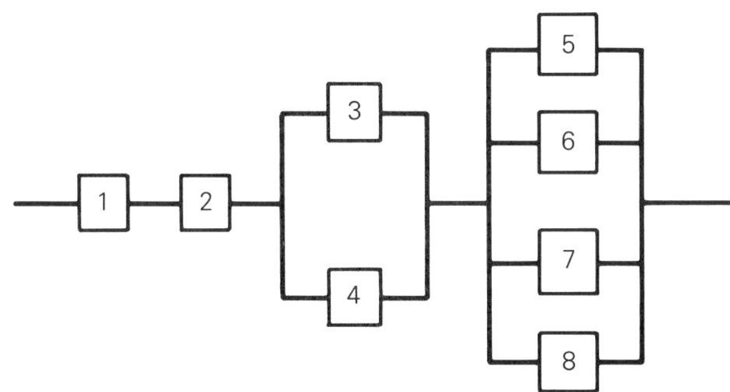

Figure 2.11 Block schematic for series parallel system.

$$\begin{aligned} R_s = {} & R_1 R_2[1 - (1 - R_3)][1 - (1 - R_4)] \\ & \times [1 - (1 - R_5)][1 - (1 - R_6)][1 - (1 - R_7)] \\ & \times [1 - (1 - R_8)] \end{aligned}$$

(ii) *Reliability scenario*

An item of equipment may operate in different modes, in different places and under different conditions at different times – for example, idle, transported to a new site, put into operation there. The calculation of its predicted reliability must take such a possibility into account. Thus if the item under consideration, at the end of a period t, will have spent times t_1, t_2 and t_3 in modes 1, 2 and 3 respectively, the reliability is

$$R(t) = R(t_1)R(t_2)R(t_3)$$

Example 6

A piece of radar equipment is placed for 2 h per day on the bridge of a ship and for 22 h on shore at sea level; if the failure rates are

on shore, sea level $\lambda_1 = 2 \times 10^{-6}$ failures per hour

on board $\lambda_2 = 3.6 \times 10^{-6}$ failures per hour

what is the reliability after 500 days?

Here t_1(on shore) = 11 000 h and t_2(on board) = 1000 h; therefore

$$\lambda_1 t_1 = 0.022 \qquad \lambda_2 t_2 = 0.0036$$

$$R(t = 500 \text{ days}) = \exp(-0.022)\exp(-0.0036) = 0.9747.$$

Example 7

A machining unit has four machines organized as in Fig. 2.12. If the separate reliabilities are

$$R_A = 0.95 \qquad R_{B1} = R_{B2} = 0.97 \qquad R_C = 0.98$$

the overall reliability is

$$R_s = (0.95)[1 - (1 - 0.97)^3](0.98) = 0.93$$

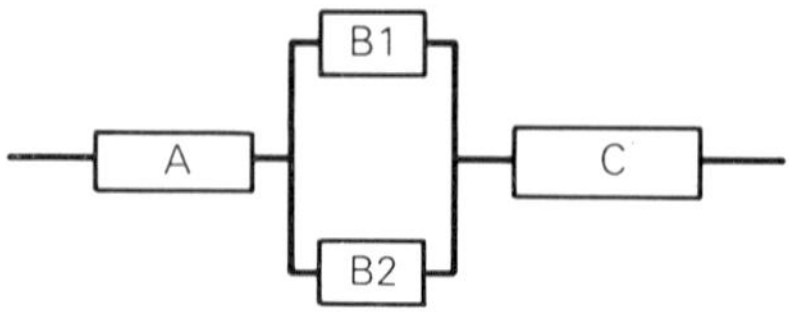

Figure 2.12 Block schematic for reliability calculation.

(iii) *Calculation of the MTBF (or MTTF)*
MTBF (mean time between failures) is relevant for repairable systems, MTTF (mean time to failure) for non-repairable systems.

The general result (see p. 11) is

$$\text{MTBF} = \int_0^\infty R(t)\,\mathrm{d}t$$

Evaluation of the integral is particularly simple when the failure rates λ_i are constant, i.e. when the exponential law $R(t) = \exp(-\lambda t)$ applies.

For a serial system, integrating the expression $R_s(t) = \Pi_i R_i(t)$ gives

$$\text{MTBF} = \frac{1}{\Sigma\,\lambda_i} = \frac{1}{n\lambda}$$

if all the λ_i are equal.

For a parallel system, integrating $R_s(t) = 1 - \Pi_i[1 - R_i(t)]$ gives

$$\text{MTBF} = \sum_i \frac{1}{\lambda_i} - \sum_{\substack{i,j\\ i\neq j}} \frac{1}{\lambda_i + \lambda_j} + \sum_{\substack{i,j,k\\ i\neq j\neq k}} \frac{1}{\lambda_i + \lambda_j + \lambda_k} - \ldots$$

If all the λ_i are equal the expression is

$$\text{MTBF} = \int_0^\infty \{1 - [1 - \exp(-\lambda t)]^n\}\,\mathrm{d}t = \frac{1}{\lambda}\left(1 + \frac{1}{2} + \frac{1}{3} + \ldots + \frac{1}{n}\right)$$

[*Translator's comment*: this result shows first that, *provided that all the elements are truly independent* (very important), the MTBF can be made as great as one wishes, because the series $1 + 1/2 + 1/3 + \ldots$ diverges; but second that there is a law of diminishing returns, and the gain from adding a further element in parallel decreases steadily. Thus the MTBF can be doubled by putting four elements in parallel, but to multiply it by 3 needs 11.]

(iv) *Interrupted tests*
There are two main types:

Sequential in which the test is stopped after some agreed number of faults, say C, have been recorded

Truncated in which the test is stopped after an agreed time, T say

Both types can be conducted with or without replacement.

Suppose that a sequential test without replacement starts at $t = 0$ with n items working and that the successive failures occur at instants $t_1, t_2, \ldots, t_c$ when c failures have been recorded and the test is stopped.

The total amount of fault-free working time is then

$$T = t_1 + t_2 + \ldots + t_c(n - c + 1)$$

and the estimate of MTBF is

$$\text{MTBF} = \frac{t_1 + t_2 + \ldots + t_c(n - c + 1)}{c}$$

In sequential testing with replacement $T = nt_c$ and MTBF $= nt_c/c$.

The truncated test without replacement finishes at a time t_f agreed in advance; suppose k failures are recorded, occurring at instants t_1, $t_2, \ldots, t_k$ $(t_k \leq t_f)$. The total good time is

$$T = t_1 + t_2 + \ldots + (n - k)t_f$$

and MTBF $= T/k$.

For the truncated test with replacement, using the same notation, $T = nt_k$ and MTBF $= T/k$.

As we shall show in Chapter 6, we can construct a confidence interval for each of these estimates. Further, the MTBF can be used to estimate the corresponding value of λ.

2.1.4 Markov chains: reliability and availability

Markov chains are a mathematical technique that enables us to compute the reliability of a system. A system consists of a set of elements connected in series and/or parallel, as in Fig. 2.13. At any instant it will be in one of a number of possible states and may or may not change to another state. We make the following basic assumptions.

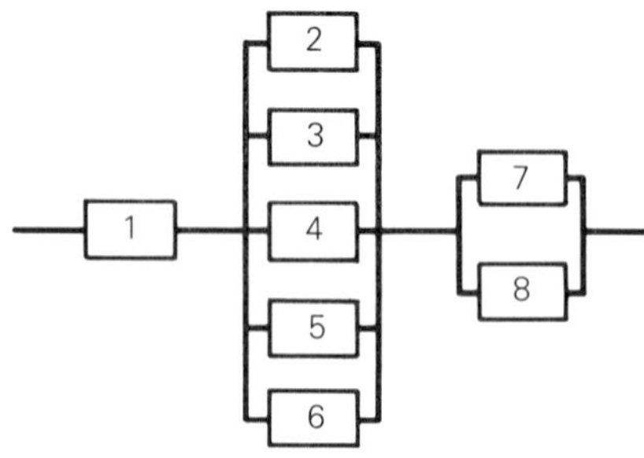

Figure 2.13 Series parallel system.

1. The possible states are numbered in such a way that when the system is in state i it can change only into $i-1$ or $i+1$; this means that the state it is in at any instant depends only on the two neighbouring states.
2. With the state changes corresponding to the failure of an element or the repair of a failed element, both failure and repair times follow an exponential law with constant rates λ and μ respectively.

(a) *Transition graph and equations*
We consider a system of n elements. We define the state i as that in which i elements are working satisfactorily; thus in state n the whole system is fault free and in state 0 it has failed completely – a breakdown. We denote the probability of changing from state i to state j by p_{ij}; it follows from (1) above that j can be only $i-1$ or $i+1$.

It is convenient (and illuminating) to represent the state changes by a labelled directed graph; Fig. 2.14 illustrates this for a three-element system. Let $P(i, t)$ be the probability that the system is in state i at time t. It will be in state i at time $t + \mathrm{d}t$ if

- it was in state i at time t and did not change during the interval $\mathrm{d}t$ or
- it was in state $i-1$ at time t and changed to i during $\mathrm{d}t$ or
- it was in state $i+1$ at time t and changed to i during $\mathrm{d}t$.

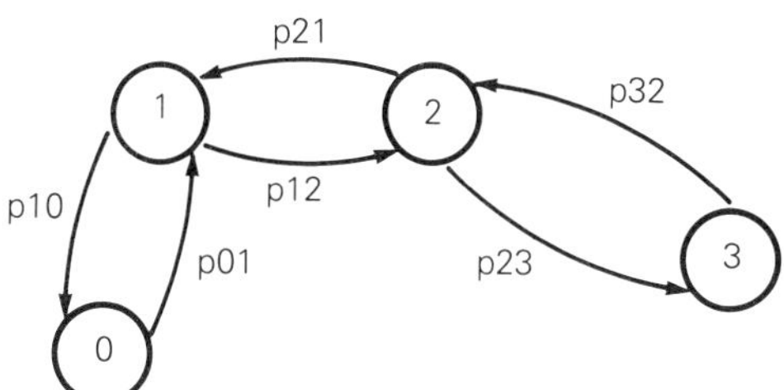

Figure 2.14 Transition graph.

These are the only possibilities; since they are independent we can add the probabilities:

$$P(i, t + \mathrm{d}t) = P(i, t)(1 - p_{i-1,i}\mathrm{d}t)(1 - p_{i+1,i}\mathrm{d}t) + P(i-1, t)p_{i-1,i}\mathrm{d}t + P(i+1, t)p_{i+1,i}\mathrm{d}t$$

Simplifying, we obtain

$$\frac{P(i, t + dt) - P(i, t)}{dt} = P(i + 1, t)p_{i+1,i} + P(i - 1, t)p_{i-1,i}$$
$$- P(i, t)(p_{i+1,i} + p_{i-1,i}) + \text{terms of order } dt$$

Letting $dt \to 0$

$$\frac{dP(i, t)}{dt} = P(i + 1, t)p_{i+1,i} + P(i - 1, t)p_{i-1,i} - P(i, t)(p_{i+1,i} + p_{i-1,i})$$

This is the most general state-change equation; assuming that the system is in full working condition at $t = 0$ the initial conditions are

$$P(i, 0) = 0 \text{ for } i \neq n \qquad P(n, 0) = 1$$

There is also the condition that at any time t the system must be in one or other of the possible states 0, 1, 2, . . ., n; thus

$$\sum_0^n P(i, t) = 1$$

i.e.

$$\sum_0^n P'(i, t) = 0$$

where $P' = dP/dt$. The equations are linear differential equations for the probabilities $P(i, t)$, and under the conditions that we have assumed the coefficients p_{ij} (the transition probabilities) are constant; they can therefore be solved by the Laplace transform method, as we shall show below.

The equations can be written very easily with the help of loops added to the transition graph: at each state i we add a loop in which we write the sum of the probabilities p_{ij} of changing to another state j with the signs reversed. Then $dP(i, t)/dt$ is equal to the sum of the products of the transition probabilities associated with each arc arriving at state i by the state from which that arc started. This is illustrated in Fig. 2.15; applying the rule gives the equation for $dP(i, t)/dt$ obtained above.

To apply these equations to any actual problem we have to know the transition probabilities p_{ij}. For the reliability problem we are studying we have

$$\text{failure rate } \lambda = \frac{1}{\text{MTBF}} \qquad \text{repair rate } \mu = \frac{1}{\text{MTTR}}$$

If there are a number of repair stations working independently the mean time to repair is reduced by that number and therefore μ is increased.

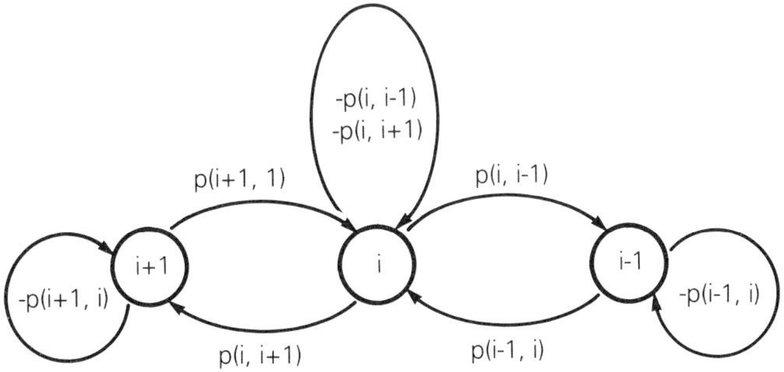

Figure 2.15 Transition graph leading to state-change differential equations.

(b) *Availability of a simple system: solution of the equations*
The simplest system consists of a single element; this element could of course be the representation of a more complex system for which we knew the failure and repair rates for the system as a whole.

There are now two and only two states, $i = 0$ or $i = 1$, with the Markov chain as in Fig. 2.16. The equations are

$$P'(1, t) = -\lambda P(1, t) + \mu P'(0, t)$$

$$P'(0, t) = \lambda P(1, t) - \mu P(0, t)$$

with

$$P(1, 0) = 1 \qquad P(0, 0) = 0$$

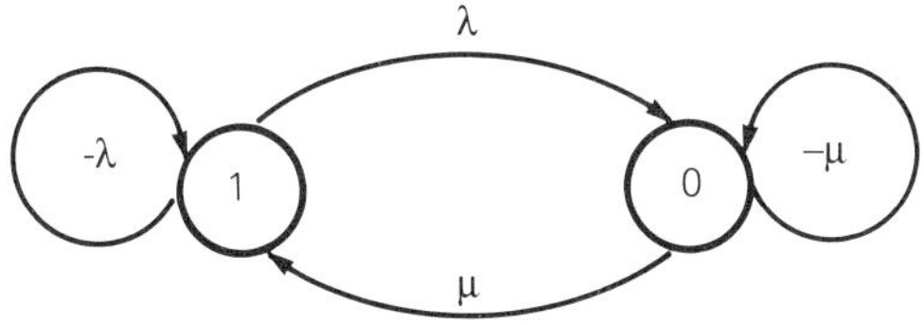

Figure 2.16 Markov chain for a two state system.

Applying the Laplace transform to these equations (L_i is the transform of $P(i, t)$)

$$pL_1 - P(1, 0) = -\lambda L_1 + \mu L_0$$

$$pL_0 - P(0, 0) = \lambda L_1 - \mu L_0$$

Putting $P(1, 0) = 1,\ P(0, 0) = 0$, we have in matrix form

$$\begin{pmatrix} p + \lambda & -\mu \\ -\lambda & p + \mu \end{pmatrix} \begin{pmatrix} L_1 \\ L_0 \end{pmatrix} = \begin{pmatrix} 1 \\ 0 \end{pmatrix}$$

We are interested only in the availability of the system, i.e. in the state in which everything is working, $P(1, t)$. Therefore we need only consider the solution for L_1:

$$L_1 = \frac{\mu + p}{p(\mu + \lambda + p)}$$

To recover $P(1, t)$ by the inverse transform $L - 1$ we must put this into partial fractions

$$L_1 = \frac{A}{p} + \frac{B}{\mu + \lambda + p}$$

We find

$$A = \frac{\mu}{\mu + \lambda} \qquad B = \frac{\lambda}{\mu + \lambda}$$

We now have

$$L_1 = \frac{A}{p} + \frac{B}{p + a}$$

where $a = \mu + \lambda$, and so from tables of the Laplace transform we find

$$P(1, t) = A + B \exp(-at)$$

$P(1, t)$, the probability that the system is in a working state, is the *availability* $A(t)$; thus we have

$$A(t) = \frac{\mu}{\mu + \lambda} + \frac{\lambda}{\mu + \lambda} \exp[-(\mu + \lambda)t]$$

As t increases, $A(t)$ tends to the constant value $\mu/(\mu + \lambda)$. Thus the result is that the availability of the system approaches the steady state value

$$A(t \to \infty) = \frac{\mu}{\mu + \lambda} = \frac{\text{MTBF}}{\text{MTBF} + \text{MTTR}}$$

(c) *System with redundant elements*

Suppose there are n identical elements, each with constant failure and repair rates λ and μ, connected in parallel; the transition graph is given in Fig. 2.17. Using the shortened notation P_i for $P(i, t)$, the equations are

$$P'_n = -n\lambda P_n + \mu P_{n-1}$$
$$P'_i = (i+1)\lambda P_{i+1} - (i\lambda + \mu) P_i + \mu P_{i-1} \qquad i = 1, 2, \ldots, n-1$$
$$P'_0 = \lambda P_1 - \mu P_0$$

with initial conditions $P_n = 1$, $P_i = 0$, $i \neq n$ at $t = 0$. These can be solved by the same method as before, but unless n is very small it is advisable to use a software package. We consider here the case $n = 2$,

$$\begin{aligned} P'_2 &= -2\lambda P_2 + \mu P_1 \\ P'_1 &= 2\lambda P_2 - (\mu + \lambda) P_1 \\ P'_0 &= \lambda P_1 - \mu P_0 \end{aligned}$$

with $P_2 = 1$ and $P_1 = P_0 = 0$ at $t = 0$.

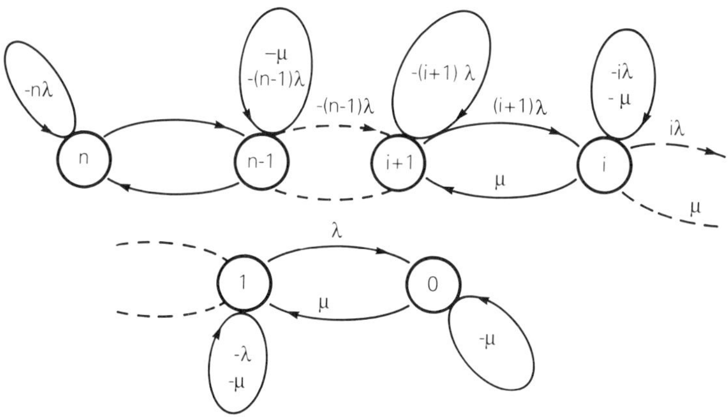

Figure 2.17 Markov chain for multi-state system.

Transforming and collecting terms as before, we obtain the equations in matrix form:

$$\begin{pmatrix} p + 2\lambda & -\mu & 0 \\ -2\lambda & p + \lambda + \mu & -\mu \\ 0 & -\lambda & p + \mu \end{pmatrix} \begin{pmatrix} L_2 \\ L_1 \\ L_0 \end{pmatrix} = \begin{pmatrix} 1 \\ 0 \\ 0 \end{pmatrix}$$

Solving, inverting and taking the constant term we find

$$A(t \to \infty) = \frac{\mu^2 + 2\mu\lambda}{\mu^2 + 2\mu\lambda + 2\lambda^2}$$

The graph of $A(t)$ as a function of t is shown in Fig. 2.18.

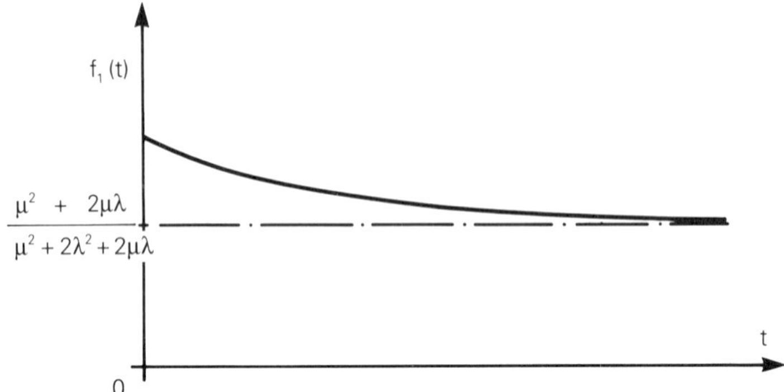

Figure 2.18 Availability function for Markov system.

Results such as these enable rational decisions to be taken on such things as the choice of technology, the amount of redundancy to build into the system, the number of repair stations, the amount of replacement stock to carry etc. It is advisable to use computer aids when dealing with complex systems.

2.1.5 Simulation: the Monte Carlo method

A simulation of a system enables a range of possibilities to be studied and hence an optimum situation to be defined. The so-called Monte Carlo method can be applied in this way to study reliability.

(a) *Principle of the method*

If the cumulative failure distribution function for the system is $F(t)$ the method is based on the idea of choosing a random sample x from a population distributed according to $F(x)$. In the reliability study x is a value for the lifetime of the element or system, say t_i for the ith sample. Suppose we draw N samples and that N_s is the number of these with $t_i > t_s$, where t_s is the required time of fault-free operation. Then

$$R(t_s) = N_s/N$$

is an estimate of the reliability at t_s.

(b) *Procedure*

There are five stages.

1. Obtain the distribution function $f(t)$ for the lifetime of the equipment under investigation.
2. Derive from this (by integration) the cumulative failure distribution function $F(t)$ (the probability that there will be at least one failure by time t).
3. Obtain a set of random numbers uniformly distributed between zero and unity.
4. Construct a random sample of lifetimes as follows: choose at random one of the numbers in (3), r say, and from the graph or table of $F(t)$ find the value of t, t_r say, such that $F(t_r) = r$. Repeat this until a sample of the required size, N say, has been drawn.
5. Use the sample to estimate the reliability, as above.

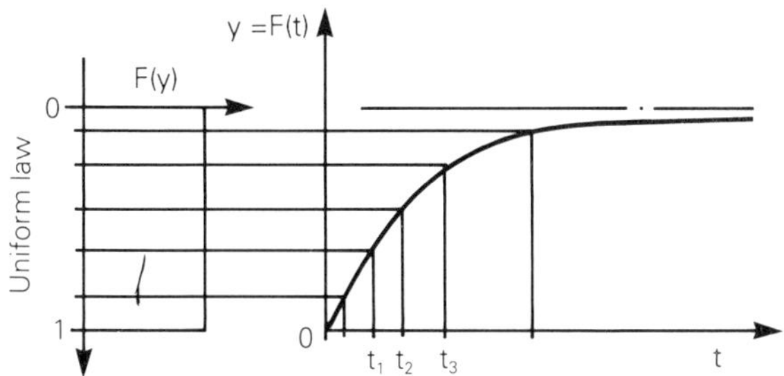

Figure 2.19 Drawing random numbers distributed between 0 and 1.

(c) *Some applications*

Example 8

The lifetime distribution for a gyroscopic system is

$$f(t) = \frac{1}{1600}\exp\left(-\frac{t}{1600}\right)$$

(t in hours). Find the reliability at $t = 2000$ h.

$$F(t) = \int_0^t f(t)\,\mathrm{d}t = 1 - \exp\left(-\frac{t}{1600}\right)$$

Ten numbers r are drawn at random from a set uniformly distributed between zero and unity, and Table 2.5 is constructed. Three of the ten trials give $t > 2000$ and so the reliability is $R(t = 2000) = 0.3$.

Table 2.5 Monte Carlo solution for Example 8

Drawing no.	$r = F(t_r)$	t_r	$t_r > 2000$
1	0.43793	927	No
2	0.07496	124	No
3	0.17405	306	No
4	0.80966	2654	Yes
5	0.65989	1725	No
6	0.55400	1292	No
7	0.72301	2054	Yes
8	0.36504	727	No
9	0.00187	3	No
10	0.90375	3745	Yes

For comparison, calculation by the exponential law gives 0.286.

There are in fact two random variables involved in a problem of this type: the lifetime of the equipment (t_r above) and the time t_s actually in service. If we draw a second set of samples to simulate the time in service we can compare the results with those for the lifetime to get another estimate of the reliability. If for convenience we rename t_r as t_1, the new estimate is

$$R(t_m) = \frac{\text{number } (t_1 > t_s)}{N}$$

Example 9

For the equipment of Example 8, suppose the time in service is distributed normally with mean 2000 h and standard deviation 150. Find the new value of the reliability.

The sampling for in-service time is done in the same way as for the lifetime, with the difference that $F(t)$ is now the cumulative normal (Gaussian) distribution with $m = 2000$, $\sigma = 150$; we obtain the required values from the table in terms of the reduced variable $u = (t - m)/\sigma$, and having found a value u we convert this back to t using $t = \sigma u + m$.

Thus if we choose the random number 0.9408 we find $u = 1.56$, i.e. $F(1.56) = 0.9408$, and so the value of t is $150 \times 1.56 + 2000 = 2234$ h. The two drawings of random numbers (RNs) are given in Table 2.6, giving again the estimate $R(t) = 3/10 = 0.3$.

This is a better procedure because it uses more values for the sample. Random numbers are easily generated by computer, and this has led to the development of software for simulation.

Table 2.6 Monte Carlo solution for Example 9

	RN(1)	t_1	RN(s)	t_s	$t_1 > t_s$?
1	0.43973	927	0.94080	2234	No
2	0.07498	124	0.27777	1911	No
3	0.17405	308	0.09621	1804	No
4	0.80968	2654	0.45577	1982	Yes
5	0.65989	1726	0.78282	2117	No
6	0.55400	1292	0.10039	1808	No
7	0.72301	2054	0.19572	1872	Yes
8	0.36504	727	0.09306	1802	No
9	0.00187	3	0.89518	2188	No
10	0.90375	3745	0.900041	2193	Yes

2.2 QUALITATIVE ANALYSIS

2.2.1 Use of failure mode analysis for quality improvement

Failure mode analysis (FMA) is a rigorous procedure for detecting potential faults, in which both the probability of the fault's occurring and the seriousness of the situation should it occur are taken into account. It is a very valuable tool for reducing the risk of equipment operating badly or failing in service and should be included in any policy for total quality control.

FMA can be applied to a product or to a process. In a manufacturing industry such as the automobile or the aeronautics industry, that uses a number of subcontractors, the main contractor should require all the suppliers and subcontractors to use FMA for products and procedures in order to guarantee quality.

For products: all possible failure modes of the system or subsystem that is being designed are noted and are taken into account in the analysis.

For procedures: the analysis takes account of all failures that can result from the manufacturing processess – assembly, casting etc.; the research and development organizations are all involved here.

2.2.2 The practice of 'Product FMA'

In implementing a policy of total quality control the analysis must be applied to

- all new components
- all components that have been modified or are to be used in new circumstances

(a) *Failure analysis*

Failure analysis is the most difficult part, and the one that demands the most skill. It involves extending the design calculations done by the research and development department to take account of all the influences that could bear on the components. Thus failures could be caused by

- deformation
- fatigue
- crack propagation
- brittle fracture
- vibration
- seizing-up
- leakages
- corrosion
- short-circuits

(b) *The aim of FMA: customer relations*

The prime importance of FMA is that it brings to light the critical issues, so that either the possibility of critical situations arising can be eliminated or means for foreseeing them can be developed. These issues will relate to certain criteria of quality, among which is the effect on the customer.

A basic rule is that the customer must not be misled. Some faults, whilst not reducing the convenience or reliability of the product, risk making a bad impression on the customer and consequently can assume great importance, particularly if they are easily detectable. They must therefore be avoided, and for this the following scales are adopted.

Probability of occurrence

4 Possible	$P > 10^{-3}$
3 Improbable	$10^{-6} < P < 10^{-3}$
2 Very improbable	$10^{-9} < P < 10^{-6}$
1 Virtually impossible	$P < 10^{-9}$

Seriousness	*Detectability*
4 Very critical	4 Very visible
3 Critical	3 Detectable
2 Not critical	2 Not very evident
1 Without effect	1 Undetectable

These will be kept in mind by those doing the analysis and are entered in the tables of results. Thus the critical issues show up and means for preventing their occurrence will appear: all this can contribute to the programme of quality improvement.

We can define a 'criticality coefficient' C as

$$C = P(\text{probability}) \times S(\text{seriousness}) \times D(\text{detectability})$$

and the most critical faults will correspond to the highest values of C.

Example 10

We consider the application of the FMA method to an adjusting device for a headlight beam – a component that would be supplied to an automobile manufacturer by subcontractors.

There are three items: a nut; an adjusting screw; an adjusting knob. Initially the manufacturer sets the adjuster with the vehicle unladen, directing the beam correctly with respect to the road (Fig. 2.20). During its life, however, the vehicle will be subjected to a variety of loadings and the device must therefore perform the function of adjusting the beam according to the load.

Taking $C \geqslant 24$ as the criterion for criticality, the analysis (Table 2.7) shows the critical issues to be as in Table 2.8.

2.2.3 Function analysis

Function analysis takes account of the relations between

- the need that has to be satisfied and the system being studied, together with its functions
- the impact of the need on the customer

Included in a study of reliability it enables

- the working of the system to be better understood
- FMA to be applied
- communication with the research and development organization to be improved
- the maintainability and cost controllability parameters to be taken into account
- failures resulting from links between different components to be brought to light in the FMA study

(a) *The method*

The system is considered as a whole and all factors relating to it are taken into account. Thus the first things to do are as follows.

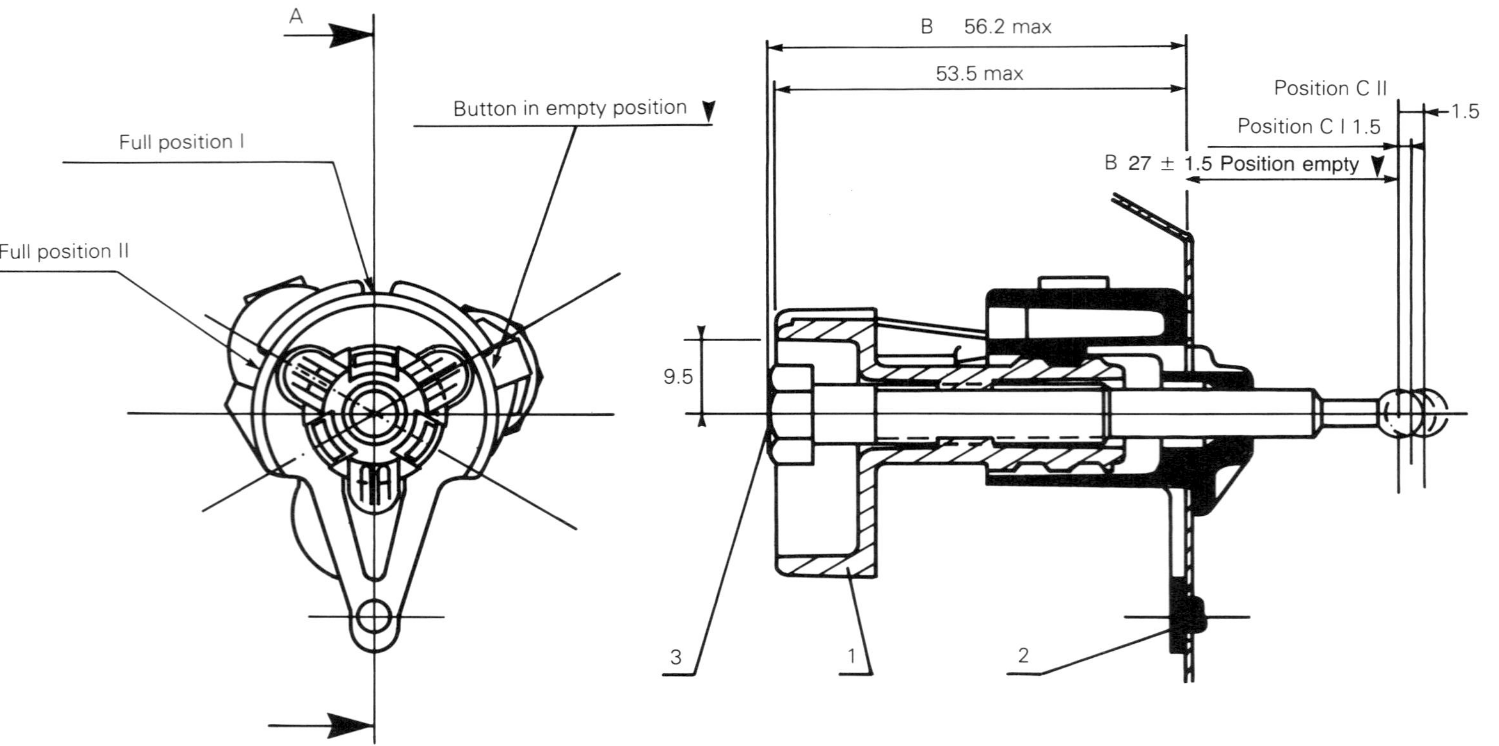

Figure 2.20 Headlight beam adjustment mechanism.

Table 2.7 Failure mode analysis (FMA) tables for headlight beam adjuster; failure modes, effects and criticality

Name	Function	Presumed mode of failure	$\lambda(t)$ (O)	Effects on equipment, subsystem, system	Symptoms observed	Means for prevention/compensation	Seriousness G	Visibility C to client	Product $G\,O\,C$	Comments and recommendations
Equipment subsystem component: adjuster knob										
Adjuster	Adjusts beam angle	Shank broken	3	Adjustment impossible	No torque	Ageing tests	4	3	36	
		Ring broken, internal diameter 7 ± 0.05	2	Adjusting screw free longitudinally, and beam angle free to vary	Screw movement irregular, beam not directed to ground	Make trials; if necessary increase ring thickness while reducing internal diameter to give constant torque	4	4	32	
		Screw thread worn/broken	1	Longitudinal freedom of adjuster	Knob movement irregular, beam not directed at ground		4	4	16	

Table 2.8 Critical failure modes and treatment for headlight beam adjuster

	Risk	Prevention
1.1	Breakage of adjusting knob	Strengthen this part
1.2	Breakage of ring	Carry out tests
3.1	Breakage of angular limit stop	Carry out accelerated fatigue and ageing tests

- Separate the functions from the hardware that realizes them.
- Identify all the functional components.

There are three main stages, concerning

- meeting the need
- defining the functions
- constructing functional block diagrams

(i) *Meeting the need*
The relevant questions are shown in Fig. 2.21. This leads to the following questions:

1. Why does this need arise?
2. How might it be eliminated?
3. What are the probabilities of the possible ways of achieving (2)?

In general, (2) will already have been considered by the customer.

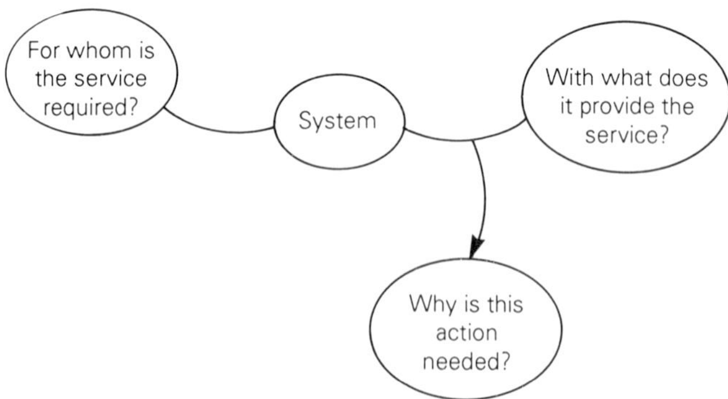

Figure 2.21 Relationship between various functions of a product or system.

(ii) *Defining the functions*
A system in a given state is in some kind of contact with its environment, to which it provides services of two types.

1. The services that are its *raison d'être*: these are its *main functions*.
2. Services that arise in response to the reactions of, or constraints applied by, the environment: these are its *response functions*.

The main functions correspond to a flow of control across the system and are therefore also called *flow functions*. There are also *design functions*: these are the elementary functions of the components and correspond to loops within the system, depending on the design.

In reliability studies the tasks to be performed are as follows.

1. Represent the functions and their interrelations by means of a functional block diagram.
2. Quantify the possibilities of breaking the flow, by *probabilities*.

(iii) *The functional block diagram*
The block diagram is a functional representation of the system in a given state of use, showing

- the external environment
- the constituent elements of the system
- the 'open' flows, i.e. the main and response flows
- the internal, design, flows
- the contacts, real or virtual, between the system elements

(b) *The different types of function of a product*
Service functions correspond to the needs expressed by the customer; they are independent of the technology employed. They must be defined with complete clarity in the specification, for on this will depend the level of satisfaction of the customer with the product delivered. They comprise the following:

- the functions that must be provided in order to meet the needs expressed;
- any additional functions (concerning aesthetics, for example) that are provided to improve the customer's opinion of the product.

Technical functions are consequential on the service functions and depend on the design. In a reliability study an analysis of the technical functions enables more to be learned about the consequences of the various failure modes.

(c) *Functional tree, functional block diagram*
The general form of the tree is of the type shown in Fig. 2.22.

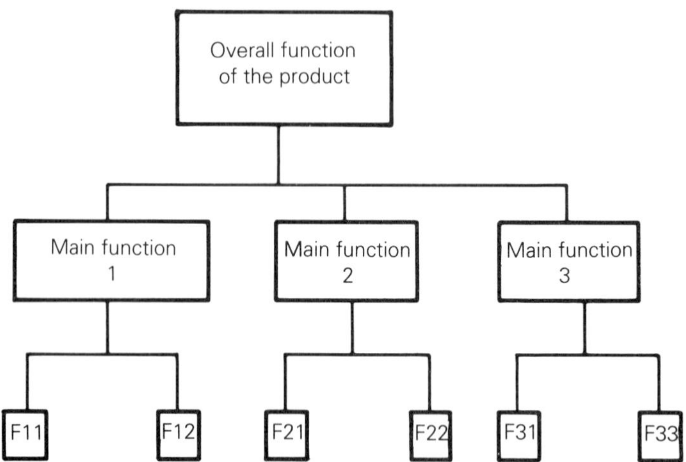

Figure 2.22 Functional tree for a typical product.

Figures 2.23 and 2.24 together with Table 2.9 show the application of this to Example 10.

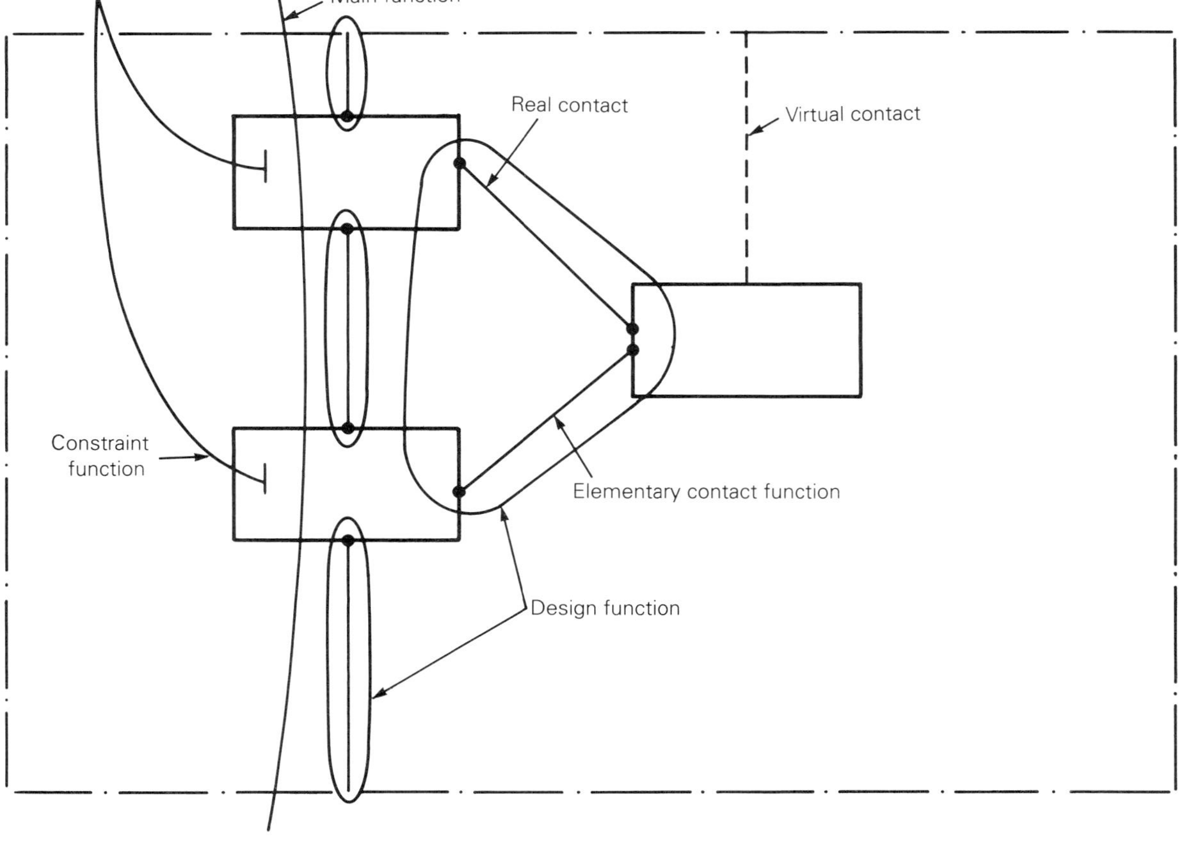

Figure 2.23 Functional relations for headlight beam adjustment.

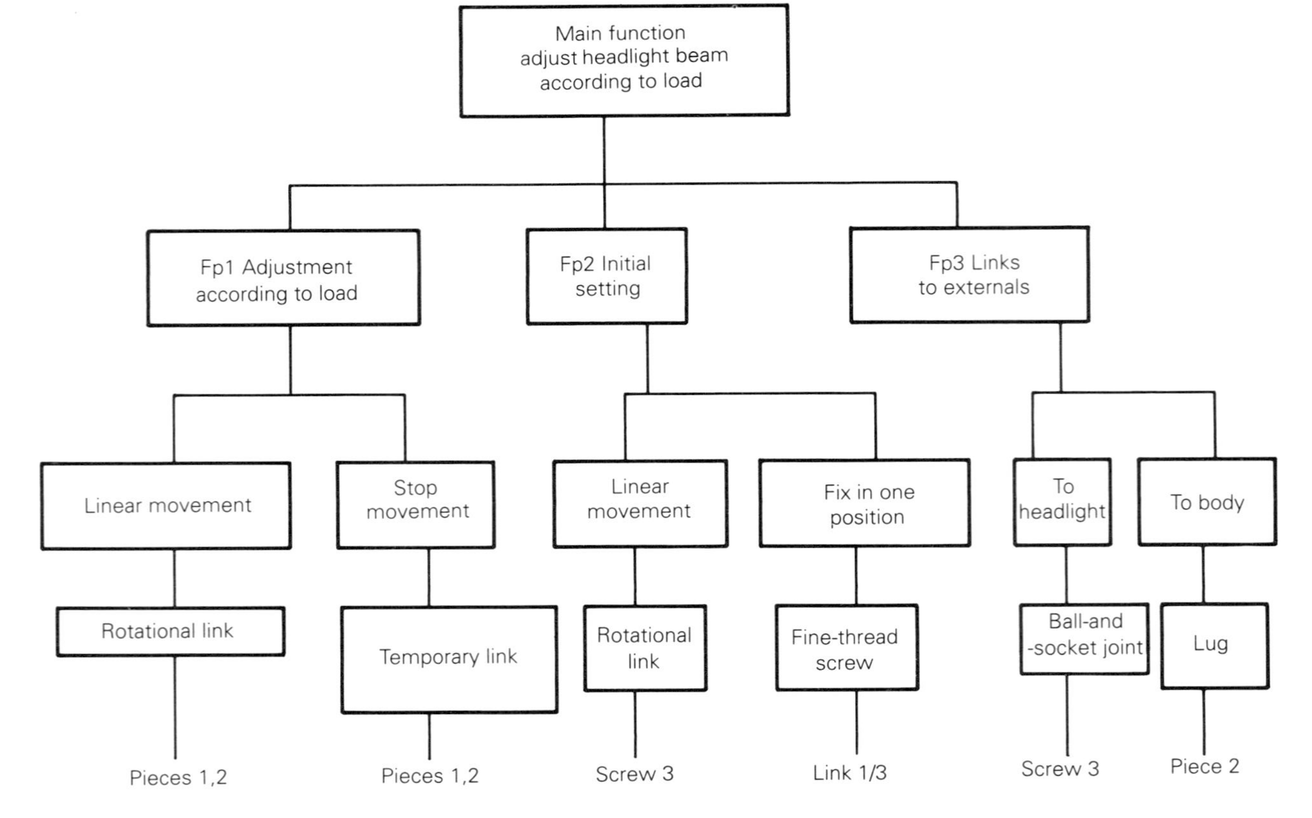

Figure 2.24 Functional tree for headlight beam adjuster.

Table 2.9 Failure mode analysis (FMA) table: components of headlight beam adjuster

Name	Function	Presumed mode of failure	$\lambda(t)$ (O)	Effects on equipment, subsystem, system	Symptoms observed	Means for prevention/ compensation	Seriousness G	Visibility C to client	Product $G\,O\,C$	Comments and recommendations
Screw thread pieces 1 and 2	Rotary link	Thread 1 broken	1	Lateral movement uncontrolled	Headlight oscillates		4	3	12	
		Thread 2 broken	1	Loss of main function (F_{p1})			4	3	12	
Link pieces 1 and 3 (claw and notch)	Prevents movement	Claw broken	2	Lateral movement not limited Loss of F_{p1}			4	4	32	
		Notch broken	1	Lateral movement not limited Loss of F_{p1}			4	3	12	

3

Controlling the manufacturing process

3.1 VARIABILITY IN MANUFACTURED PRODUCTS

Manufactured products that are supposed to be identical will in fact vary, and it is very important to understand this variability; such an understanding can lead to

- reduction in the number of items to be scrapped,
- better adaptation of the machines to the production programme,
- better appreciation of the problems of the production processes in the production planning office and
- better use of the control charts.

The study we give here is conducted in terms of an example from mechanical engineering (Fig. 3.1); however, the concepts apply equally to the manufacture of electronic components and indeed to any type of serial production.

We consider a very simple item, a spacer, manufactured in quantity, where a certain dimension is specified; but when samples from a batch are measured with a precision gauge the values found are as given in Table 3.1.

These variations can result from a variety of factors:

- temperature changes
- vibration
- positioning in the machine tool
- deformation
- flexure of the machining tool
- wear of the machining tool

They can be put into two classes:

1. random variations (RV)—these can be of either sign, and it is not possible to predict which.
2. systematic variations (SV)—the development of these can be predicted.

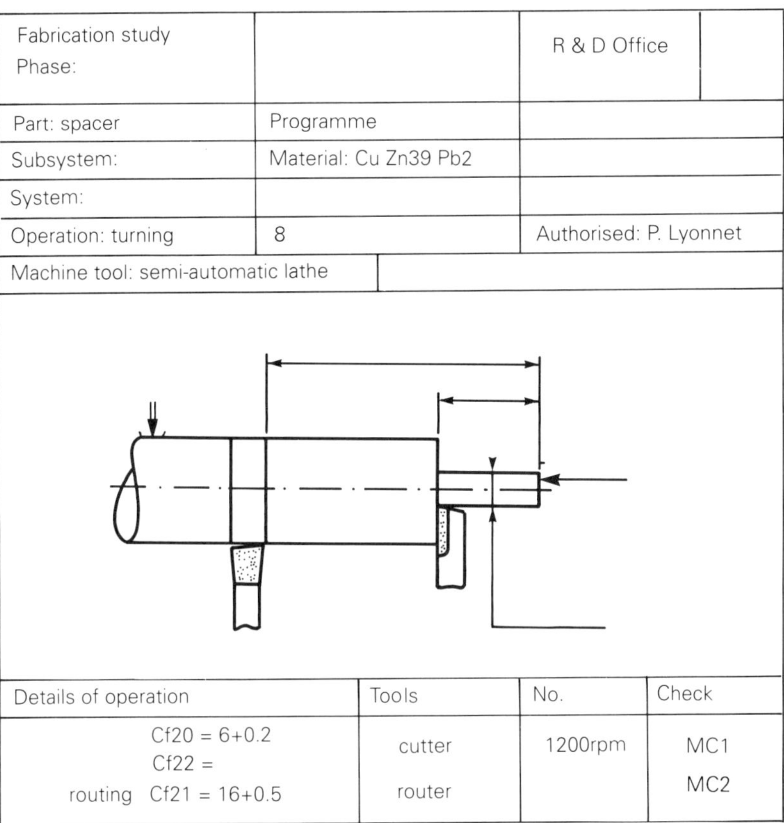

Fabrication study Phase:		R & D Office
Part: spacer	Programme	
Subsystem:	Material: Cu Zn39 Pb2	
System:		
Operation: turning	8	Authorised: P. Lyonnet
Machine tool: semi-automatic lathe		

Details of operation	Tools	No.	Check
Cf20 = 6+0.2 Cf22 = routing Cf21 = 16+0.5	cutter router	1200rpm	MC1 MC2

Figure 3.1 Machining instructions.

Table 3.1 Sample of spacing pieces

No.	Values	No.	Values	No.	Values
1	5.95	1	5.96	1	5.93
2	5.94	2	5.96	2	6.12
3	6.00	3	5.93	3	6.06
4	5.94	4	5.98	4	6.01
5	6.00	5	5.98	5	5.94
6	5.99	6	6.09	6	5.82
7	5.97	7	5.96	6	5.82
7	5.97	7	5.96	7	6.07
8	6.10	8	5.92	8	5.92
9	5.98	9	5.93	9	5.97
10	5.94	10	5.94	10	5.94
X	5.981	X	5.971	X	5.978
W	0.16	W	0.17	W	0.3

X = mean, W = range.

The total variation TV is the sum of these: TV = RV + SV.

It is important to distinguish between the two types and to measure both. If a group of items is taken from a batch, measurements on these will show the random variations but will not reveal anything about wear or drift.

3.1.1 Random variations: an example

After the volume production process has settled down a sample of 20 is measured, with results as in Table 3.2. When plotted, these give the histogram in Fig. 3.2. The form is characteristic of a random variation; the curve of the normal (Gaussian) law is superimposed for comparison.

Table 3.2 Measured values for 20 samples from a production run

No.	Value	No.	Value	No.	Value	No.	Value
1	6.10	6	5.80	11	6.20	16	6.00
2	6.05	7	5.86	12	5.92	17	5.94
3	5.98	8	5.90	13	6.04	18	6.10
4	6.00	9	6.01	14	5.92	19	5.98
5	6.00	10	6.05	15	6.01	20	5.95

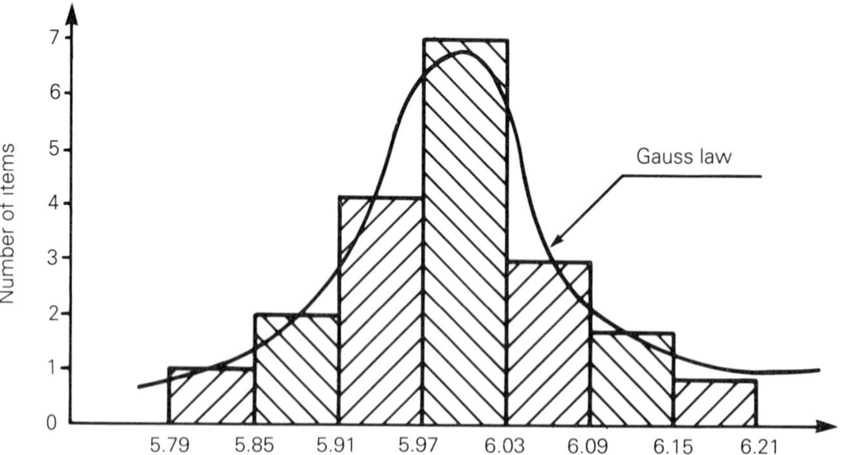

Figure 3.2 Histogram showing distribution of values for data in Table 3.2.

We shall assume generally that random variations follow the normal law: in any particular application this can be checked by applying a goodness-of-fit test such as the χ^2 or Kolmogorov–Smirnov test. If the parameters of the distribution are the mean m and the standard deviation σ we can then use the facts that about 95% of the values will lie in the range $m \pm 2\sigma$ and about 99% in $m \pm 3\sigma$. We can find estimates for m and σ from a sample. The mean for the sample is an estimate for m, and if W is the range or spread of values in the sample, i.e. the difference between the greatest and least values, an estimate for σ is W/d_n, where n is the number of items in the sample and d_n is a known function (Table 3.3 gives values of d_n).

Table 3.3 Estimation of σ from range $W : \hat{\sigma} = W/\mathrm{d_n}$

Size of each sample	$1/d_n$	d_n
2	0.886	1.128
3	0.591	1.693
4	0.486	2.059
5	0.430	2.326
6	0.395	2.534
7	0.370	2.704
8	0.351	2.847
9	0.337	2.970
10	0.325	3.078
11	0.315	3.173
12	0.307	3.258

Another important property of the normal law is that the mean of a sample of size n is itself distributed normally, with mean m and standard deviation $\sigma/n^{1/2}$, i.e. the sample mean is the same as the mean for a single item but its range of variation is narrower by a factor $n^{1/2}$.

3.1.2 Systematic variation: an example

(a) *Introduction*
Ten samples of the same item are measured (a) at the start of a production run, (b) after 80 have been made and (c) after 160 have been produced. The results are given in Table 3.4.

Table 3.4 Samples taken at different stages showing systematic change of the mean (x)

1st sample start of series		2nd sample after 80 items		3rd sample after 160 items	
No.	Values	No.	Values	No.	Values
1	6.08	1	5.85	1	5.61
2	5.94	2	5.72	2	5.68
3	6.06	3	5.86	3	5.61
4	6.00	4	5.73	4	5.65
5	6.08	5	5.80	5	5.55
6	5.93	6	5.90	6	5.68
7	6.09	7	5.71	7	5.52
8	5.92	8	5.85	8	5.62
9	5.98	9	5.72	9	5.51
10	5.93	10	5.82	10	5.69
X	6.00	X	5.79	X	5.61
W	0.17	W	0.17	W	0.18

Systematic variation is indicated by a *steadily changing mean*. In this example we see that the mean X is decreasing with the number of items machined, whilst the range W (the spread about the mean) remains effectively constant. This indicates wear of the machine tool, the development of which can be represented by a straight line.

In section 3.2 we shall show that this linear variation of wear can be used to construct a control chart for the process.

(b) *An application*

Let TI be the tolerance interval for a measurement of an item; then if the variations are random with standard deviation σ we must have

$$\mathrm{TI} \geqslant 6\sigma$$

Otherwise there is a risk that items are rejected.

Example 1

For the piece being considered, suppose that the dimension C is given by $C = 5.00 \pm 0.20$, so that $\mathrm{TI} = 0.40$. A sample of 10 gives the following measurements:

5.10, 4.95, 4.93, 5.12, 5.15, 4.90, 5.10, 4.95, 4.85, 5.04

The range $W = 5.15 - 4.85 = 0.30$; therefore an estimate for σ (using Table 3.3) is

$$\sigma = \frac{0.30}{d_n} = \frac{0.30}{3.078} = 0.097$$

This gives $6\sigma = 0.6$, which is greater than the tolerance 0.4. Therefore we can expect that items will be rejected.

Example 2

Suppose that $C = 6.00 \pm 0.025$ and TI = 0.50. A sample of 10 gives the following:

5.84, 5.90, 6.10, 6.00, 5.97, 5.90, 6.05, 5.98, 6.10, 6.05

$w = 6.10 - 5.84 = 0.26$; thus $\sigma = 0.26/3.078 = 0.084$ and $6\sigma = 0.5$. Hence TI $= 6\sigma$, which is a warning that the process should be monitored.

Example 3

$C = 4.00 \pm 0.30$ and sample values are as follows:

4.10, 4.05, 3.95, 4.05, 3.97, 4.03, 3.95, 4.02, 3.96, 4.04

$w = 4.10 - 3.95 = 0.15$, $\sigma = 0.15/3.078 = 0.048$, $6\sigma = 0.29$ and TI $> 6\sigma$. Therefore there should be no risk of rejects.

3.2 MONITORING THE MANUFACTURE

The aim of checking during manufacture is to keep the manufacturing process under control and hence to ensure uniform production, whether the items manufactured are electronic or mechanical components, food products or anything else. To achieve this we must be able to detect any deviation from the norm so that we can make the necessary adjustments before the process produces items that have to be rejected.

The parameters monitored can be either attributes – and classed as either 'good' or 'bad' – or properties that can be measured; measurements of properties give more effective control but are not always available. The monitoring uses a graphical presentation called *control charts*; these can be shown to the customer to justify any claim to quality of production.

3.2.1 Control charts using measurements

The essential requirement is that the parameter used is a measurable property, for example length, weight, electrical resistance. As already stated, we assume here that the variations observed have a normal, or Gaussian, distribution, and so we can use the general results of section 3.1.1.

When the tolerances on the measurement are known a control chart does the following:

- it shows up any drift in the measurement concerned;
- it enables the intervals at which adjustments should be made to be calculated;.
- it enables the need for a major resetting to be foreseen;
- it shows up any increase in the range of variation of the measurement concerned, and therefore the need to examine the machine;
- it enables the quality of the manufacture to be assessed.

If the tolerances are not known the chart provides only the first four of these.

The chart is very much a picture of the manufacture and highlights any problems clearly; this is illustrated by the examples in Fig. 3.3. When the parameter plotted on the chart is the mean or the median the range or the standard deviation should be shown also, to give an overall picture of the process.

(a) *Control chart for the mean*

(i) *Mean and standard deviation known*

The chart is a plot of the mean values of the parameter we wish to control for a series of samples of the same size, n say. We know that if the value of this parameter for a single item has a normal distribution with mean m and standard deviation σ the sample mean will have a normal distribution with mean m and standard deviation $\sigma/n^{1/2}$. We therefore mark on the chart (Fig. 3.4)

$$\text{upper/lower control limits } m \pm 3.09\sigma/n^{1/2}$$

$$\text{upper/lower monitoring limits } m \pm 1.96\sigma/n^{1/2}$$

95% of the points should lie within the monitoring limits and 99.8% within the control limits. Thus if a newly plotted point lies on or just beyond one of the monitoring limits this is a warning that the process needs watching; if it lies on or beyond a control limit some corrective action is needed.

(ii) *Mean and standard deviation not known*

When the mean and standard deviation are not known we have to estimate these parameters of the law. Suppose that we have r samples each of size n and let m_i, σ_i be the mean and standard deviation respectively of the values for the ith sample. The estimates are as follows:

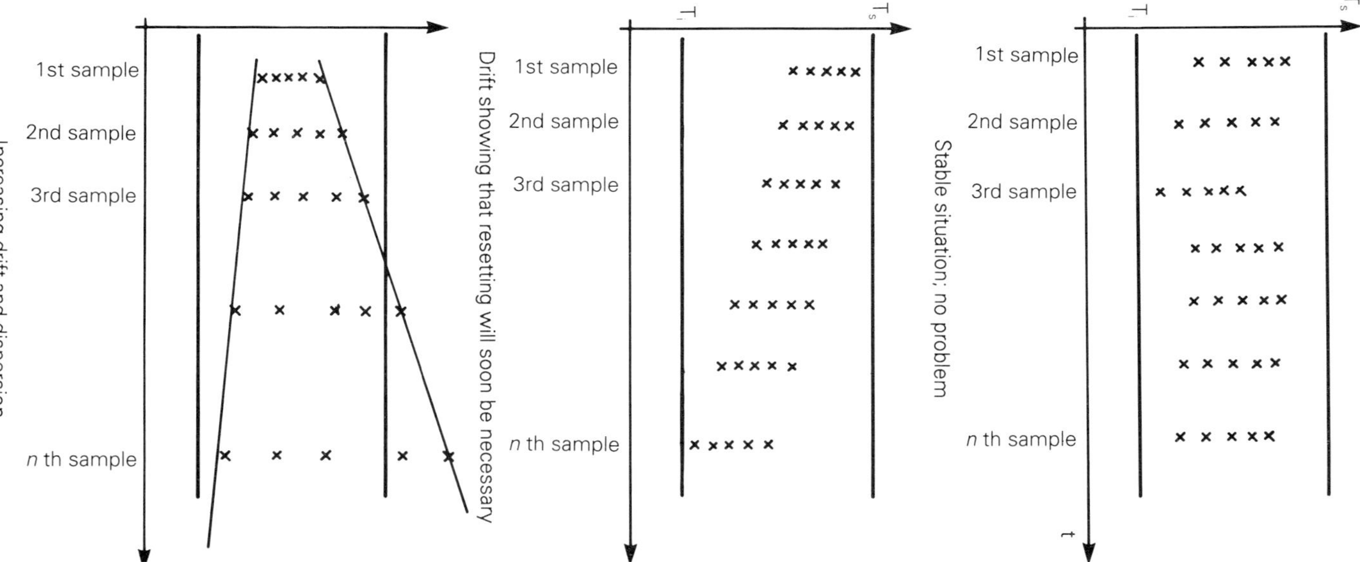

Figure 3.3 Typical loss of control in a production run.

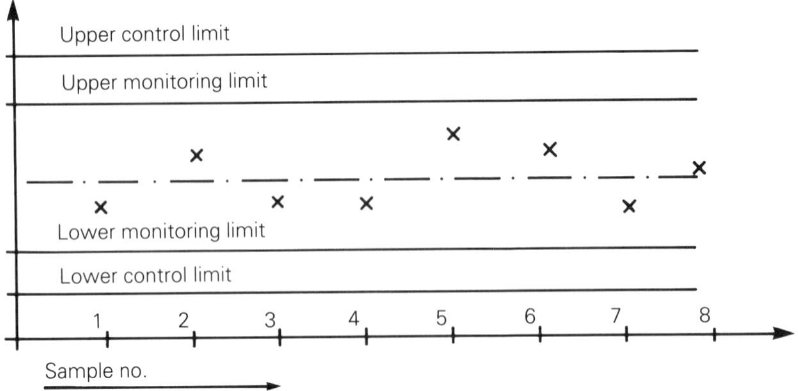

Figure 3.4 Control chart: Mean and standard deviation not known.

$$m = \frac{\Sigma m_i}{r}$$

$$\sigma = \frac{1}{b_n} \frac{\Sigma \sigma_i}{r}$$

where b_n is given in Table 3.5. If x_{ij}, $j = 1, 2, \ldots, n$ are the values measured in sample i then

$$m_i = \frac{\Sigma x_{ij}}{n} \qquad \sigma_i = \left[\frac{\Sigma_j (x_{ij} - m_i)^2}{n - 1}\right]^{1/2}$$

The denominator $n - 1$, instead of n, in the expression for σ_i gives what is called an unbiased estimate for the sample standard deviation.

For $n < 12$ an estimate for σ_i is W_i/d_n, where W_i is the range for the sample i and d_n is the function of n referred to in section 3.1.2(a); values are given in Table 3.3. This is quicker, but less accurate.

When these estimates for m and σ have been found they can be used as above to construct the control chart.

Control charts of this kind are used when the tolerance limits are not known, or when the spread of values is very narrow.

If the tolerance interval is known and is large compared with the standard deviation, the limits on the chart must be modified to give more flexibility in the manufacture. This is discussed in the next section.

Example 4

Four samples of a product, each of five items, are measured, with the result given in Table 3.6, from which we find

m_i	15.247	15.244	15.238	15.254
σ_i	0.018	0.015	0.008	0.011

Table 3.5 Estimation of population standard deviation from sample value: coefficient b

Size of each sample	$1/b_n$	b_n
2	1.773	0.564
3	1.381	0.724
4	1.253	0.798
5	1.189	0.841
6	1.151	0.869
7	1.126	0.888
8	1.107	0.903
9	1.094	0.914
10	1.083	0.923
11	1.075	0.930
12	1.068	0.936
13	1.063	0.941
14	1.058	0.945
15	1.054	0.949
16	1.050	0.952
17	1.047	0.955
18	1.044	0.958
19	1.042	0.960
20	1.040	0.962
21	1.037	0.964
22	1.035	0.966
23	1.034	0.967
24	1.033	0.968
25	1.031	0.970
26	1.030	0.971
27	1.029	0.972
28	1.028	0.973
29	1.027	0.974
30	1.026	0.975

$$\dot{\hat{\sigma}}_0 = \frac{1}{b_n} \bar{\sigma}_n$$

and $$\sigma_n = \left[\frac{1}{n} \sum_i^n (X_i - \bar{X})^2\right]^{1/2}$$

$$\hat{\sigma}_r = \frac{1}{r} \sum_{j=1}^{r} \sigma_{n_j}$$

Table 3.6 Measurements of samples from a production run

Sample 1	Sample 2	Sample 3	Sample 4
15.250	15.260	15.240	15.240
15.220	15.230	15.230	15.250
15.240	15.240	15.250	15.270
15.260	15.260	15.230	15.260
15.265	15.230	15.240	15.250

The means of these are 15.244 and 0.013 respectively; thus the estimates for the population mean and standard deviation are

$$m = 15.244 \qquad \sigma = \frac{0.013}{b_n} = 0.013 \times 1.189 = 0.015$$

Hence

$$\text{upper/lower control limits} = 15.244 \pm 3.09 \times 0.015/\sqrt{5} = 15.244 \pm 0.021$$
$$= 15.265,\ 15.223$$
$$\text{upper/lower monitor limits} = 15.244 \pm 1.96 \times 0.015/\sqrt{5} = 15.244 \pm 0.013$$
$$= 15.257,\ 15.233$$

from which the control chart can be constructed.

(b) *Modification to take account of known tolerance limits*
If the tolerances are known and the interval is large compared with the standard deviation the chart should be modified to take these into account, in order to avoid unnecessary adjustments to the process.

In Fig. 3.5, T_u and T_l are the upper and lower tolerance limits respectively, i.e. the limits between which the measurement *must* lie; lines A are modified control limits and B are modified monitoring limits. It can be shown that

$$\overline{T_u - A} = 3.09\sigma - \frac{3.09\sigma}{n^{1/2}} = \overline{T_l - A}$$

$$\overline{T_u - B} = 3.09\sigma - \frac{1.96\sigma}{n^{1/2}} = \overline{T_l - B}$$

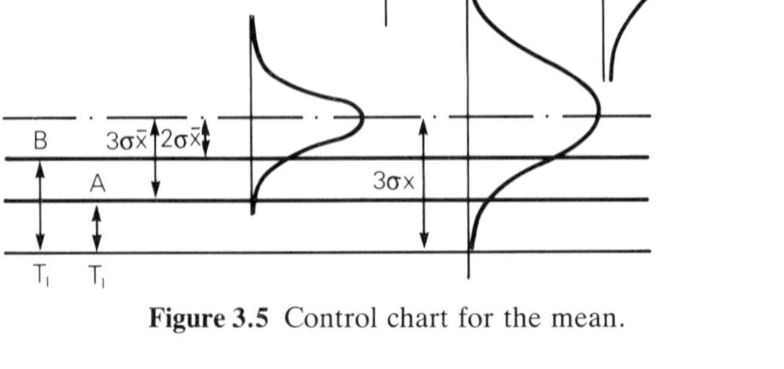

Figure 3.5 Control chart for the mean.

If the value of the population standard deviation σ is not known, an estimate can be found as described above.

Example 5
Suppose the tolerance limits are $T_u = 122.350$ and $T_l = 122.000$ and that the estimate for σ from samples of five specimens is 0.016.

$$3.09\sigma - \frac{3.09\sigma}{\sqrt{5}} = 0.027 \qquad 3.09\sigma - \frac{1.96\sigma}{\sqrt{5}} = 0.035$$

from which it follows that the modified limits are as follows.

Control: 122.350 − 0.027 = 122.323   122.000 + 0.027 = 122.027

Monitor: 122.350 − 0.035 = 122.315   122.000 + 0.035 = 122.035

(c) *Control charts for the variations*
The derivation of the control chart for the mean is based on the assumption that the variation is stable, and it is therefore necessary to ensure that this is so. For this we need to control

- the standard deviation (control chart for σ)
- the spread (control chart for W)

(i) *Control chart for standard deviation*
We suppose that the population standard deviation σ is known; when this is so it can be shown that, if $\hat{\sigma}$ is the estimate obtained from a sample, the quantity $n\hat{\sigma}/\sigma$ where n is the size of the sample is distributed as χ^2 with $n - 1$ degrees of freedom. Then if C_u, C_l, M_u and M_l are the upper and lower limits for control and monitoring respectively

$$C_u = B_{cu}\sigma$$

and correpondingly for the other limits, where

$$B_{cu} = \left[\frac{\chi^2(0.999;\ n - 1)}{n}\right] \qquad B_{cl} = \left[\frac{\chi^2(0.001;\ n - 1)}{n}\right]$$

$$B_{mu} = \left[\frac{\chi^2(0.975;\ n - 1)}{n}\right] \qquad B_{ml} = \left[\frac{\chi^2(0.025;\ n - 1)}{n}\right]$$

Values of the coefficients B as functions of the sample size n are given in Table 3.7.

If σ is not known these limits are given in terms of the estimated standard deviation $\hat{\sigma}$ by a set of relations

$$C_u = B'_{cu}\hat{\sigma}_0 \text{ etc.}$$

and the modified coefficients B' are given in Table 3.8.

Table 3.7 Determination of control chart limits: standard deviation known

Size of each sample	B_{cu}	B_{cl}	B_{mu}	B_{ml}
2	0.001	2.327	0.022	1.585
3	0.026	2.146	0.130	1.568
4	0.078	2.017	0.232	1.529
5	0.135	1.922	0.311	1.493
6	0.187	1.849	0.372	1.462
7	0.223	1.791	0.420	1.437
8	0.274	1.744	0.459	1.415
9	0.309	1.704	0.492	1.396
10	0.339	1.670	0.520	1.379
11	0.367	1.640	0.543	1.365
12	0.391	1.614	0.564	1.352
13	0.413	1.591	0.582	1.340
14	0.432	1.570	0.598	1.329
15	0.450	1.552	0.613	1.320
16	0.467	1.535	0.626	1.311
17	0.482	1.520	0.637	1.303
18	0.495	1.505	0.648	1.295
19	0.508	1.492	0.658	1.288
20	0.520	1.480	0.667	1.282
21	0.531	1.469	0.676	1.276
22	0.541	1.458	0.684	1.270
23	0.551	1.449	0.691	1.265
24	0.560	1.439	0.698	1.260
25	0.569	1.431	0.704	1.255
26	0.577	1.423	0.710	1.250
27	0.584	1.415	0.716	1.246
28	0.592	1.408	0.721	1.242
29	0.599	1.401	0.727	1.238
30	0.605	1.394	0.731	1.235

Example 6

If σ(known) = 0.012 and the sample size $n = 5$, we have

$$C_u = 1.922 \times 0.012 = 0.023 \qquad C_l = 0.135 \times 0.012 = 0.002$$

$$M_u = 1.493 \times 0.012 = 0.018 \qquad M_l = 0.311 \times 0.012 = 0.004$$

Table 3.8 Determination of control chart limits: standard deviation unknown

Size of each sample	B'_{cu}	B'_{cl}	B'_{mu}	B'_{ml}
2	0.002	4.126	0.039	2.810
3	0.036	2.964	0.180	2.166
4	0.098	2.528	0.291	1.916
5	0.161	2.285	0.370	1.775
6	0.215	2.128	0.428	1.682
7	0.262	2.017	0.473	1.618
8	0.303	1.931	0.508	1.567
9	0.338	1.864	0.538	1.527
10	0.367	1.809	0.563	1.494
11	0.395	1.763	0.584	1.468
12	0.418	1.724	0.603	1.444
13	0.439	1.691	0.618	1.424
14	0.457	1.661	0.633	1.406
15	0.474	1.635	0.646	1.391
16	0.491	1.612	0.658	1.377
17	0.505	1.592	0.667	1.364
18	0.517	1.571	0.676	1.352
19	0.529	1.554	0.685	1.342
20	0.541	1.538	0.693	1.333
21	0.551	1.524	0.701	1.324
22	0.560	1.509	0.708	1.315
23	0.570	1.498	0.715	1.308
24	0.579	1.487	0.721	1.302
25	0.587	1.475	0.726	1.294
26	0.594	1.465	0.731	1.287
27	0.601	1.456	0.737	1.282
28	0.608	1.447	0.741	1.276
29	0.615	1.438	0.746	1.271
30	0.621	1.430	0.750	1.267

An important point to note here is that if the sample standard deviation falls to one of the lower limits, or below, this is not a signal to intervene in the manufacturing process but rather to inspect the measuring equipment so as to ensure that it is not giving faulty readings. A real fall in the standard deviation – i.e. in the spread – is relatively rare.

(ii) *Control chart for range*

There are corresponding coefficients D, D' for the cases when the population standard deviation σ is and is not known respectively. When σ is known

$$C_u = D_{cu}\sigma \quad \text{etc.}$$

and when σ is not known

$$C_u = D'_{cu}W \quad \text{etc.}$$

where W is the sample range or the mean of the ranges of a number of samples. Values of D, D' are given in Tables 3.9 and 3.10.

Table 3.9 Control chart for range: standard deviation known

Size of each sample	D_{cu}	D_{cl}	D_{mu}	D_{ml}
2	0.00	4.65	0.04	3.17
3	0.06	5.06	0.30	3.68
4	0.20	5.31	0.59	3.98
5	0.37	5.48	0.85	4.20
6	0.54	5.62	1.06	4.36
7	0.69	5.73	1.25	4.49
8	0.83	5.82	1.41	4.61
9	0.96	5.90	1.55	4.70
10	1.08	5.97	1.67	4.79
11	1.20	6.04	1.78	4.86
12	1.30	6.09	1.88	4.92

Table 3.10 Control chart for range: standard deviation unknown

Size of each sample	D'_{cu}	D'_{cl}	D'_{mu}	D'_{ml}
2	0.00	4.12	0.04	2.81
3	0.04	2.99	0.18	2.17
4	0.10	2.58	0.29	1.93
5	0.16	2.36	0.37	1.81
6	0.21	2.22	0.42	1.72
7	0.26	2.12	0.46	1.66
8	0.29	2.04	0.50	1.62
9	0.32	1.99	0.52	1.58
10	0.35	1.94	0.54	1.56
11	0.38	1.90	0.56	1.53
12	0.40	1.87	0.58	1.51

Example 7

If the range $W = 0.122$ for a sample of size $n = 10$ the four coefficients D'_{cu} etc. are 1.94, 0.35, 1.56, 0.54; multiplying the observed range 0.122 by these gives the limits

$$\text{control} \quad (0.234, 0.043)$$

$$\text{monitor} \quad (0.195, 0.065)$$

For practical use of the chart it is convenient to shade the bands between the control and monitor limits, as in the example in Fig. 3.6. In the case of the chart for the standard deviation only the upper limits need be shown.

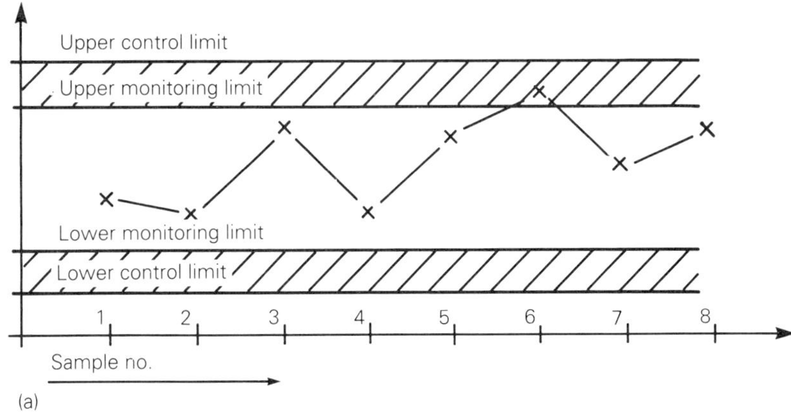

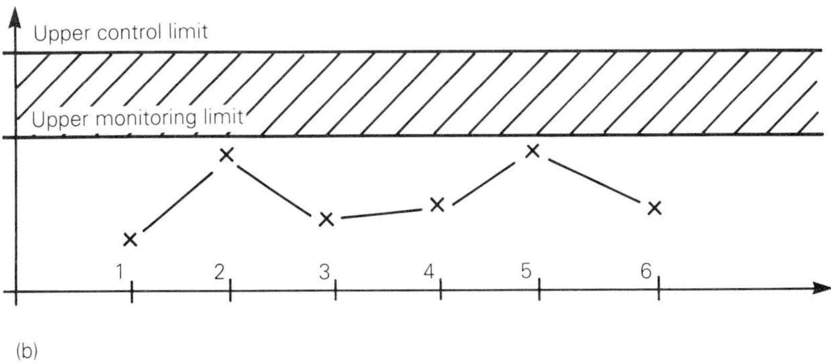

Figure 3.6 (a) Control chart for the mean (example 7). (b) Control chart for the standard deviation (example 7).

(d) *Control chart with directly plotted individual values*
With this form no calculation at all is done, the individual values being plotted directly (Fig. 3.7). The rule here is that the occurrence either of two successive values in a shaded band or of one value outside either control limit is a signal to take action.

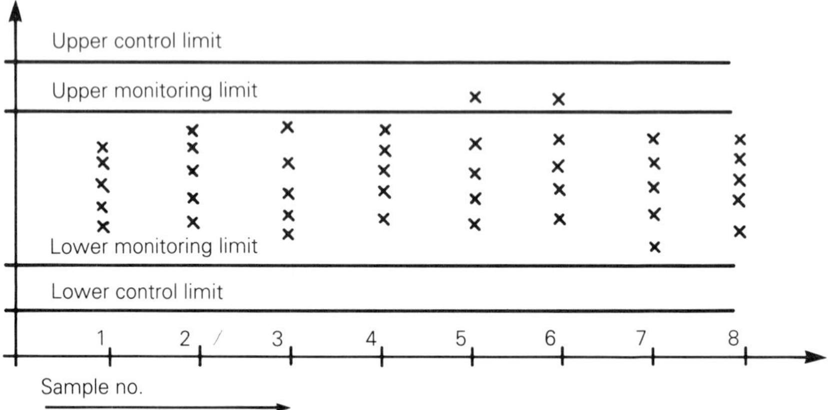

Figure 3.7 Control chart for individual values.

In constructing this form of chart the assumption is made that the tolerance interval TI is six times the population standard deviation: $TI = 6\sigma$. Thus the estimate for σ is $\hat{\sigma} = TI/6$. Further, the distances d_1 and d_2 of Fig. 3.8 can be found using $d_1 = P_1 \times TI$ and $d_2 = P_2 \times TI$ where P_1 and P_2 are found from Table 3.11.

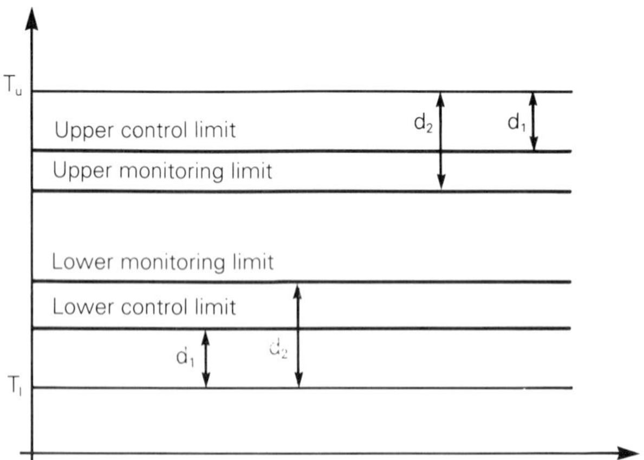

Figure 3.8 Control chart for the number of defects allowed.

Table 3.11 Determination of limits for Figure 3.8

n		$P = 1\%$ $(U = 2.6)$	$P = 0.5\%$ $(U = 2.8)$	$P = 0.27\%$ $(U = 3)$
3	P_1	2.6%	6.5%	9.3%
	P_2	16.3%	19.1%	21.1%
4	P_1	+ 0.7%	4.7%	7.6%
	P_2	13.4%	16.4%	18.5%
5	P_1	− 0.8%	3.4%	6.4%
	P_2	11.2%	14.4%	16.7%
6	P_1	− 1.9%	2.3%	5.4%
	P_2	9.6%	12.9%	15.3%
7	P_1	− 3.0%	1.4%	4.5%
	P_2	8.3%	11.7%	14.2%
8	P_1	− 3.8%	0.6%	3.8%
	P_2	7.2%	10.7%	13.2%
9	P_1	− 4.5%	− 0.1%	3.2%
	P_2	6.3%	9.9%	12.4%
10	P_1	− 5.1%	− 0.6%	2.6%
	P_2	5.4%	9.1%	11.7%

Example 8
Find the limits for the individual value chart with

$$T_u = 150.250 \qquad T_l = 150.000 \qquad \mathrm{TI} = 0.250$$

$$n = 6$$

and the percentage P of defectives acceptable equal to 0.5.

For these values Table 3.11 gives $d_1 = 0.023\mathrm{TI}$, $d_2 = 0.129\mathrm{TI}$; hence

$$C_u = 150.250 - 0.023 \times 0.250 = 150.244$$

$$M_u = 150.250 - 0.129 \times 0.250 = 150.218$$

$$M_l = 150.000 + 0.129 \times 0.250 = 150.032$$

$$C_l = 150.000 + 0.023 \times 0.250 = 150.006$$

3.2.2 Control charts for attributes

Not all the properties that affect quality can be measured: this is especially the case where any kind of aesthetic judgement is involved, as for example with the coachwork of an automobile. The control criteria must then be qualitative and expressed in terms such as ‘good/bad’, although of course some quantitative control may be involved, such as some dimension being or not being within the allowable limits. In these circumstances it is useful to distinguish between control charts for

- the number of defective items
- the proportion of defective items
- the number of defects per item

(a) *Control chart for the number of defectives in a sample, the proportion in the population being known*
The control and monitor limits are defined as those within which 99.8% and 95% respectively of the population lie; their values are calculated on the assumption that the fraction P of defectives in the population is known.

If the sample is of n items and the total population is N items then (see Chapter 6) if $n/N < 0.1$ the binomial law can be used to calculate the probability of any number of defectives in the sample; otherwise the hypergeometric law must be used.

Assuming that the binomial law is applicable, the upper control limit C_u is defined as follows (we are not, of course, interested in a lower limit):

$$\sum_{k=0}^{C_u} \binom{n}{k} P^k(1 - P)^{n-k} = 0.998$$

The upper monitor limit is given by

$$\sum_{k=0}^{M_u} \binom{n}{k} P^k(1 - P)^{n-k} = 0.975$$

These limits are now integers as we are dealing with discrete variables; consequently we may not be able to achieve the probability value 0.998 and 0.975 exactly.

Values for the limits can be found by using Tables 3.12 and 3.13 which have been computed from the binomial and Poisson distributions.

Example 9
The proportion of defectives is stable at 3%; we want to know the control and monitor limits for a sample size of 15.

Looking down the columns headed $n = 15$ of Table 3.13 we see that the nearest we can get to 3% is with

$C_u = 3$, corresponding to $P = 3.1\%$

$M_u = 1$, corresponding to $P = 1.9\%$ ($M_u = 2$ corresponds to 4.3%)

Table 3.12 Data for the calculation of control chart limits for numbers and proportions of defective items

	Upper control and monitor limits[a]							
C_u or M_u	m_0 or np_0 for C_u	m_0 or np_0 for M_u	C_u or M_u	m_0 or np_0 for C_u	m_0 or np_0 for M_u	C_u or M_u	m_0 or np_0 for C_u	m_0 or np_0 for M_u
0	0.001	0.025	10	3.49	5.49	20	9.62	13.00
1	0.045	0.24	11	4.04	6.20	21	10.29	13.79
2	0.19	0.62	12	4.61	6.92	22	10.96	14.58
3	0.43	1.09	13	5.20	7.65	23	11.65	15.38
4	0.74	1.62	14	5.79	8.40	24	12.34	16.18
5	1.11	2.20	15	6.41	9.15	25	13.03	16.98
6	1.52	2.81	16	7.03	9.90	26	13.73	17.79
7	1.97	3.45	17	7.66	10.67	27	14.44	18.61
8	2.45	4.12	18	8.31	11.44	28	15.15	19.42
9	2.96	4.80	19	8.96	12.22	29	15.87	20.24

[a] Based on Poisson approximation to the binomial distribution

Table 3.13 Table for computing control chart limits L_u, L_1 for the number of proportion of defectives per sample

C_u or M_u	$n = 10$		$n = 15$		$n = 20$		$n = 30$		$n = 40$		$n = 50$		C_u or M_u
	p_0% for C_u	p_0% for M_u	p_0% for C_u	p_0% for M_u	p_0% for C_u	p_0% for M_u	p_0% for C_u	p_0% for M_u	p_0% for C_u	p_0% for M_u	p_0% for C_u	p_0% for M_u	
0	0.01	0.25	0.01	0.17		0.13		0.09		0.06		0.05	0
1	0.5	2.5	0.3	1.9	0.23	1.3	0.15	0.8	0.11	0.6	0.09	0.5	1
2	2.1	6.7	1.4	4.3	1.0	3.2	0.66	2.1	0.5	1.7	0.4	1.4	2
3	4.9	12.2	3.1	7.8	2.2	5.7	1.5	3.7	1.1	2.7	0.9	2.2	3
4	8.8	18.7	5.5	11.8	4.0	8.7	2.5	5.6	1.9	4.2	1.3	3.3	4
5	14.1	26.2	8.5	16.3	6.1	11.9	3.9	7.7	2.9	5.6	2.2	4.5	5
6	20.4	34.8	12.1	21.3	8.6	15.4	5.4	9.9	4.0	7.3	3.1	5.8	6
7	28.1	44.4	16.1	26.6	11.3	19.1	7.2	12.3	5.2	9.0	4.2	7.1	7
8	37.6	55.5	20.6	32.3	14.4	23.1	9.0	14.8	6.5	10.8	5.2	8.5	8
9	50.1	69.2	25.7	38.4	17.7	27.2	11.0	17.3	8.0	12.7	6.2	10.0	9
10			31.3	44.9	21.3	31.5	13.1	19.9	9.5	14.6	7.4	11.5	10
11			37.5	51.9	25.1	36.1	15.3	22.6	11.1	16.5	8.6	13.1	11
12			44.6	59.5	29.3	40.8	17.7	25.4	12.7	18.5	10.0	14.6	12
13			52.8	68.1	33.7	45.7	20.2	28.3	14.4	20.6	11.3	16.2	13
14			63.0	78.2	38.4	50.9	22.7	31.3	16.2	22.7	12.6	17.8	14
15					43.5	56.3	25.4	34.3	18.1	24.8	14.0	19.5	15
16					49.1	62.1	28.2	37.4	20.0	27.0	15.5	21.2	16
17					55.2	68.3	31.1	40.6	22.0	29.2	17.0	22.9	17
18					62.3	75.1	34.1	43.9	24.0	31.5	18.5	24.6	18
19					70.8	83.2	37.2	47.3	26.0	33.8	20.1	26.4	19
20							40.4	51.9	28.1	36.1	21.6	28.2	20

p_0 not specified, $n \leqslant 50$.
p_0 is given as a percentage, C_u and M_u as numbers of defectives.

(b) *Control chart for the proportion of defectives*
A control chart for the proportion of defectives is preferable when the size of the sample can vary. For a sample of n the corresponding limits are

$$C'_{u} = \frac{C_{u}}{n} \qquad M'_{u} = \frac{M_{u}}{n}$$

and if as before k is the number of defectives in the sample the quantity to be controlled is

$$p = k/n$$

Thus for the above example, $C'_{u} = 3/15 = 0.2$, $M'_{u} = 1/15 = 0.067$.

3.3 INTERVAL BETWEEN CONTROL ACTIONS

One of the problems facing the quality engineer is to predict the intervals at which control actions should be taken; this will depend on the way in which the equipment can get out of adjustment, which can be

- by random changes or
- by systematic drifts, due to wear, overheating etc.

3.3.1 Random variations

A study of the production statistics will enable the desirable interval between control actions to be estimated; the control charts will indicate when the process is going out of control. However, it is important to find the reasons for these changes, and a study of the correlations between the changes and the parameters of the process will help here.

3.3.2 Drifts

This type of change is easier to correct because, given enough observations, regression analysis (linear, polynomial, exponential or logarithmic) can be used to model the law describing the change.

Tool wear is a cause of drift in mechanical engineering production, and tribological studies have shown that this is practically linear over the active life of the tool: this is illustrated in Fig. 3.9. Thus regression can be used to model the change.

Tool life can be expressed as a distance, the distance travelled either by the tool itself or the metal or other material that is being machined. If y is the amount of wear (also measured as a distance) when the tool life is x, the relation assumed is

$$y = ax + b$$

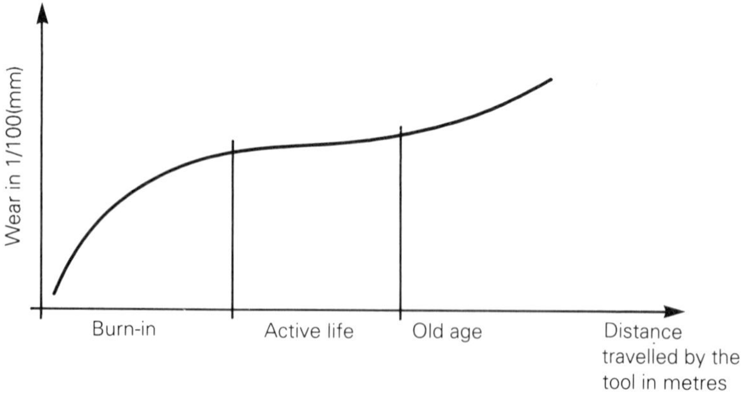

Figure 3.9 Typical life history for machine tool wear.

If we have n pairs of observations (x_i, y_i), with means X and Y respectively, the least squares fitting of the straight line gives (Fig. 3.10)

$$a = \frac{\Sigma\, x_i y_i - nXY}{\Sigma\, (x_i)^2 - nX^2}$$

$$b = \bar{Y} - a\bar{X}$$

This can now be used to predict the stage at which the wear will have reached the maximum that can be allowed.

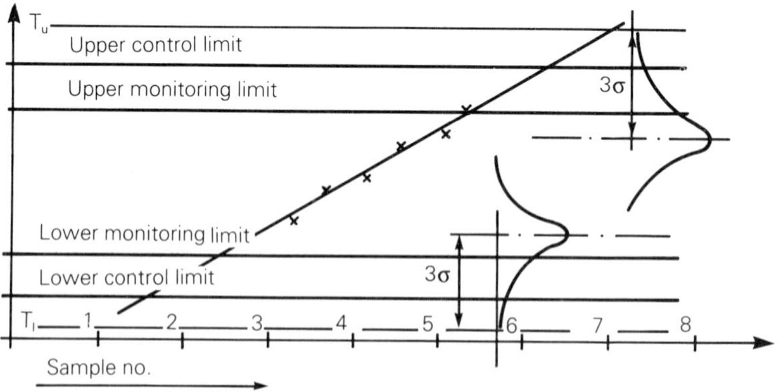

Figure 3.10 Control chart with successive samplings showing cumulative effect of wear.

Example 10
Find the regression line for the following record of values, and the tool life (in metres) if the allowable wear is 2 mm.

Here y (in millimetres) is the relevant dimension of the piece machined when the tool life is x (in metres).

y_i	35.10	35.15	35.16	35.17	35.19	35.20	35.21	35.24	35.30	35.40
x_i	500	505	506	508	509	511	511.5	515	520	530.5

Using the results given above we find the line to be

$$y = 30.162 + 9.877(x/1000)$$

Thus a change Δx results in a change Δy where

$$\Delta y = 9.877(\Delta x/1000)$$

and so, for $\Delta y = 2$, $\Delta x = 2000/9.877 = 202$ m.

Example 11 involving the control chart
We wish to establish a control routine, with chart, for an item to which the following apply: tolerances $T_1 = 75.50$, $T_u = 76.50$; sample size $n = 5$. Measurements on a pre-production sample give

75.90 75.93 75.90 75.92 75.92 75.95 75.93 75.94

75.99 75.96 75.93 75.96 75.97 75.98 75.99

Measurements made in order to determine tool wear are given in Table 3.14, where y_i is the dimension for the ith piece machined. The time required to machine one piece is 12 min.

Table 3.14 Data for example 11

i	y_i	i	y_i
1	75.000	42	75.042
3	75.003	80	75.080
10	75.009	90	75.081
25	75.026	100	75.101
30	75.031	150	75.149
35	75.034		

We first estimate the population standard deviation σ: using the formula given in section 3.2.1(a) with the values for the pre-production sample we find

$$\hat{\sigma} = 0.0294$$

Next, we find the limits for the control chart for the mean by using the method of section 3.2.1(b):

$$T_u A = 3\hat{\sigma} - \frac{3\hat{\sigma}}{\sqrt{5}} = 0.0489 = T_l A \qquad T_u B = 3\hat{\sigma} - \frac{2\hat{\sigma}}{\sqrt{5}} = 0.062 = T_l$$

The chart limits are therefore

$$C_u = T_u - T_u A = 76.500 - 0.049 = 76.451$$

$$M_u = T_u - T_u B = 76.500 - 0.062 = 76.438$$

$$M_l = T_l + T_u B = 75.500 + 0.062 = 75.562$$

$$C_l = T_l + T_u A = 75.500 + 0.049 = 75.549$$

Next we find the chart limits for σ, for which only the upper values are of interest:

$$C_u = B_{cu}\hat{\sigma} = 1.922 \times 0.0294 = 0.056$$

$$M_u = B_{mu}\hat{\sigma} = 1.493 \times 0.0294 = 0.044$$

(from Table 3.7). We can now construct the control charts.

To determine the frequency of control action, we first find the amount of wear that can be allowed:

$$M = \mathrm{TI} - 6\sigma = (76.500 - 75.500) - 6 \times 0.0294 = 0.824$$

The regression line for the wear data is found to be (Fig. 3.11)

$$y = 9.85 \times 10^{-4} + 74.999$$

and so the number of pieces machined when the wear has reached 0.824 is $0.824/9.85 \times 10^{-4} = 836$.

Finally, since the time required is 12 min per piece, the tool should be reset at intervals of $836 \times 12/60$ h = 167 h of working life.

```
10  REM REGRESSION LINE
20  CLS
30  LOCATE 8,15: INPUT "How many (X,Y) pairs do you wish to consider";
    N
40  CLS
50  LOCATE 4,51: PRINT "X        Y"
60  DIM X(N): DIM Y(N)
70  CLS
80  FOR I=1 TO N
90  LOCATE I*2, 5: PRINT "PAIR";I;: INPUT "Value of X=    ", X(I)
```

Figure 3.11 BASIC program for finding regression lines: only X_i, Y_i pairs need be entered.

```
100  LOCATE I*2,40: INPUT “Value of Y=     ”, Y(I)
110  NEXT I
120  CLS
130  LOCATE 4,6: PRINT “Number of pairs (X,Y):”; PRINT N
140  LOCATE 4,51: PRINT “X        Y”
150  FOR I=1 TO N
160  LOCATE 5+I,50: PRINT X(I), Y(I)
170  LOCATE 5+I,45: PRINT CHR$(179)
180  PRINT: X=X+X(I): Y=Y+Y(I): X2=X2+X(I)^2:
     Y2=Y2+Y(I)^2: XY=XY+X(I)*Y(I)
190  NEXT I
200  LOCATE 6,1
210  PRINT “Sum of Xs              :”; X
220  PRINT “Sum of Ys              :”; Y
230  MX=X/N: MY=Y/N
240  PRINT “   Mean (X)             :”; MX
250  PRINT “   Mean (Y)             :”; MY
260  Print “Sum of squares (X)         :”; X2
270  Print “Sum of squares (Y)         :”; Y2
280  MXY=XY/N
290  VX=X2/N−MX^2: VY=Y2/N−MY^2
300  PRINT “Variance X            :”; VX
310  PRINT “Variance Y            :”; VY
320  COV=XY/N−(MX*MY)
330  PRINT “Covariance (X,Y)          :”; COV
340  RO=COV/(SQR(VX)*SQR(VY))
350  PRINT USING “   Correlation coefficient      :#.###”;RO
360  PRINT “Regression line”
370  PRINT “   ----------   ”
380  A=COV/VX: B=MY−R*MX
390  PRINT USING “   Coefficient A : #.###”; A
400  PRINT USING “   Coefficient B : #.###”; B
410  FOR I=1 TO 19
420  LOCATE 2+I,1: PRINT CHR$(186)
430  LOCATE 2+I,71: PRINT CHR$(186)
440  NEXT I
450  LOCATE 22,1: PRINT CHR$(200)
460  LOCATE 2,1: PRINT CHR$(201)
470  FOR I=1 TO 70
480  LOCATE 22,I+1: PRINT CHR$(205)
490  LOCATE 2,I+1: PRINT CHR$(205)
500  NEXT I
510  LOCATE 2,71: PRINT CHR$(187)
520  LOCATE 22,71: PRINT CHR$(188)
530  LOCATE 23,10: INPUT “Any more? (Y/N)”; A$
540  IF A$=Y THEN 20 ELSE 550
550  END
```

Figure 3.11 (continued)

4

Quality control of goods received

Quality control of goods received is an important component of any quality programme; its purpose is to filter out any below-standard items or materials delivered to a part of the enterprise

- from outside, by any of its suppliers, or
- from inside, by another unit or department.

The ways in which control is usually exercised are

- by attribute(s): the item concerned is rated 'good' or 'bad' according to some criterion, and as with control charts the decision on whether to accept or reject the delivery is based on the number of 'bad' items in the sample.
- by number of defects per item: the mean number is found, by sampling, and the decision to accept or reject is based on this.
- by measurements: the property on which the decision is to be based must be measurable, as when used for a control chart; the sample mean and standard deviation can be used as criteria.

4.1 CONTROL BY ATTRIBUTES

4.1.1 Types of sampling

(a) *Simple sampling*
A single sample is taken and the decision whether to accept or reject the batch is based on the number of defectives found.

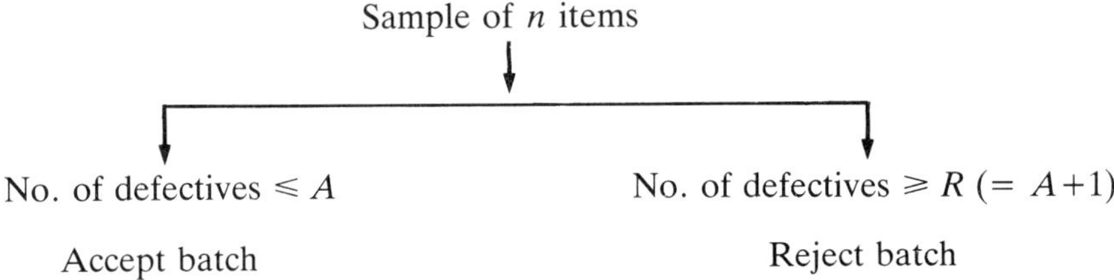

Here A is the acceptance criterion, R the rejection criterion.

This is the simplest control procedure to implement; however, it is not optimal from an economic point of view, and for batches that are definitely good or definitely bad the decision can be based on a smaller sample, thus reducing the cost. This can be achieved by double or multiple sampling.

(b) *Double sampling*
A first sample is taken as before; depending on the result either a decision is reached or a second sample is taken and the decision is based on the result of the two combined.

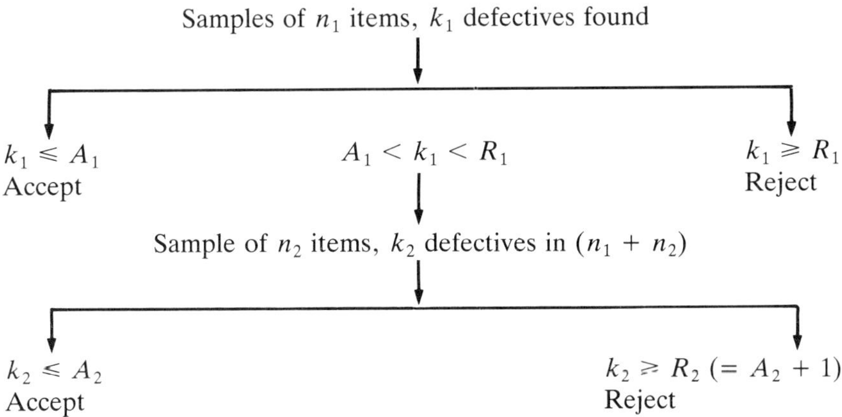

(c) *Multiple sampling*
This process can be repeated: the criteria for the second sample are

	$k_2 \leqslant A_2$	accept
	$k_2 \geqslant R_2$	reject $(R_2 \neq A_2 + 1)$
$A_2 <$	$k_2 < R_2$	take a third sample n_3:
		k_3 defectives in $n_1 + n_2 + n_3$
	$k_3 \leqslant A_3$	accept
	$k_3 \geqslant R_3$	reject $(R_3 = A_3 + 1)$

and so on, up to eight samples (this number is defined by AFNOR). The values of the criteria A_1, R_1 etc. are found by using the standard statistical distributions – binomial, hypergeometric, Poisson.

4.1.2 Laying down a control procedure

The calculations to define any of a number of control procedures have been done and the results are given in the French standard NF 022 X; this greatly simplifies the task of implementing the procedure in any particular case. Alternatively, a software package can be used.

The following have to be specified for the procedure:

- the type of control to be used (by attributes or by measured properties)
- the sample size
- the method of sampling
- the relation between the result of the sampling and the decision taken

The *type of control* will depend on whether the property of interest is or is not measurable.

The *sample size* will depend on the size of the batch and on the stringency with which the control is to be applied; Tables 4.1 and 4.2 give standard recommendations, in which

- levels S1, S2, S3 and S4 apply to military applications only,
- level I is for relatively relaxed control, Level II is normal and Level III is for relatively strict control.

Through the code letter given in Table 4.1, Table 4.2 gives the values of n_1 for simple sampling, n_2 for double and n_i for higher multiplicities.

The *method of sampling* will have to be agreed by the two parties.

For each sampling scheme the *decision criteria* are functions of the acceptable quality level (AQL), expressed as the fraction or percentage of defective items in the batch (see Fig. 4.5 later).

4.1.3 Risks borne by the supplier and by the customer

It is important to realize that there are inherent risks of making the wrong decision in any control system based on sampling, and these must be accepted by the two parties.

Suppose the true (unknown) fraction of defectives in the batch is p, and that the batch is acceptable if $p \leqslant p_1$ and not acceptable if $p \geqslant p_2$. If k is the number of defectives found in a sample of n items, then with the notation of section 4.1.1 this means $A = np_1$, $R = np_2$. The risks are as follows:

Table 4.1 Letter code for determining sample size

Lot or batch size	Special inspection levels				General inspection levels		
	S_1	S_2	S_3	S_4	I	II	III
2 to 8	A	A	A	A	A	A	B
9 to 15	A	A	A	A	A	B	C
16 to 25	A	A	B	B	B	C	D
26 to 50	A	B	B	C	C	D	E
51 to 90	B	B	C	C	C	E	F
91 to 150	B	B	C	D	D	F	G
151 to 280	B	C	D	E	E	G	H
281 to 500	B	C	D	E	F	H	J
501 to 1 200	C	C	E	F	G	J	K
1 201 to 3 200	C	D	E	G	H	K	L
3 201 to 10 000	C	D	F	G	J	L	M
10 001 to 35 000	C	D	F	H	K	M	N
35 001 to 150 000	D	E	G	J	L	N	P
150 001 to 500 000	D	E	G	J	L	N	P
500 001 and over	D	E	H	K	N	Q	R

Table 4.2(a) Simple sampling plans with normal control

Code Letter	Sample Size	Acceptable Quality Level (normal control)																									
		0.010	0.015	0.025	0.040	0.065	0.10	0.15	0.25	0.40	0.65	1.0	1.5	2.5	4.0	6.5	10	15	25	40	65	100	150	250	400	650	1000
		A R	A R	A R	A R	A R	A R	A R	A R	A R	A R	A R	A R	A R	A R	A R	A R	A R	A R	A R	A R	A R	A R	A R	A R	A R	A R
A	2															0 1			1 2	2 3	3 4	5 6	7 8	10 11	14 15	21 22	30 31
B	3														0 1			1 2	2 3	3 4	5 6	7 8	10 11	14 15	21 22	30 31	44 45
C	5													0 1			1 2	2 3	3 4	5 6	7 8	10 11	14 15	21 22	30 31	44 45	
D	8												0 1			1 2	2 3	3 4	5 6	7 8	10 11	14 15	21 22	30 31	44 45		
E	13											0 1			1 2	2 3	3 4	5 6	7 8	10 11	14 15	21 22	30 31	44 45			
F	20										0 1			1 2	2 3	3 4	5 6	7 8	10 11	14 15	21 22						
G	32									0 1			1 2	2 3	3 4	5 6	7 8	10 11	14 15	21 22							
H	50								0 1			1 2	2 3	3 4	5 6	7 8	10 11	14 15	21 22								
J	80							0 1			1 2	2 3	3 4	5 6	7 8	10 11	14 15	21 22									
K	125						0 1			1 2	2 3	3 4	5 6	7 8	10 11	14 15	21 22										
L	200					0 1			1 2	2 3	3 4	5 6	7 8	10 11	14 15	21 22											
M	315				0 1			1 2	2 3	3 4	5 6	7 8	10 11	14 15	21 22												
N	500			0 1			1 2	2 3	3 4	5 6	7 8	10 11	14 15	21 22													
P	800		0 1			1 2	2 3	3 4	5 6	7 8	10 11	14 15	21 22														
Q	1250	0 1			1 2	2 3	3 4	5 6	7 8	10 11	14 15	21 22															
R	2000			1 2	2 3	3 4	5 6	7 8	10 11	14 15	21 22																

Table 4.2(b) Multiple sampling plans with normal control

Sample size code letter	Number of samples	Sample size	Accumulated sample size	Acceptable Quality Level (normal control)																									
				0.010	0.015	0.025	0.040	0.065	0.10	0.15	0.25	0.40	0.65	1.0	1.5	2.5	4.0	6.5	10	15	25	40	65	100	150	250	400	650	1000
				A R	A R	A R	A R	A R	A R	A R	A R	A R	A R	A R	A R	A R	A R	A R	A R	A R	A R	A R	A R	A R	A R	A R	A R	A R	A R
A B C																*	*	*			*	*	*	*	*	*	*	*	*
D	First	2	2												*			* 2	* 2	* 3	* 4	0 4	0 5	1 7	2 9	4 12	6 16		
	Second	2	4															* 2	0 3	0 3	1 5	1 6	3 8	4 10	7 14	11 19	17 27		
	Third	2	6															0 2	0 3	1 4	2 6	3 8	6 10	8 13	13 19	19 27	29 39		
	Fourth	2	8															0 3	1 4	2 5	3 7	5 10	8 13	12 17	19 25	27 34	40 49		
	Fifth	2	10															1 3	2 4	3 6	4 8	7 11	11 15	17 20	25 29	36 40	53 58		
	Sixth	2	12															1 3	3 5	4 6	5 9	10 12	14 17	21 21	31 33	45 47	65 68		
	Seventh	2	14															2 3	4 5	6 7	6 10	13 14	18 19	25 26	37 38	53 54	77 78		
E	First	3	3											*			* 2	* 2	* 3	* 4	0 4	0 5	1 7	2 9	4 12	6 16			
	Second	3	6														* 2	0 3	0 3	1 5	1 6	3 8	4 10	7 14	11 19	17 27			
	Third	3	9														0 2	0 3	1 4	2 6	3 8	6 10	8 13	13 19	19 27	29 39			
	Fourth	3	12														0 3	1 4	2 5	3 7	5 10	8 13	12 17	19 25	27 34	40 49			
	Fifth	3	15														1 3	2 4	3 6	4 8	7 11	11 15	17 20	25 29	36 40	53 58			
	Sixth	3	18														1 3	3 5	4 6	5 9	10 12	14 17	21 21	31 33	45 47	65 68			
	Seventh	3	21														2 3	4 5	6 7	6 10	13 14	18 19	25 26	37 38	53 54	77 78			
F	First	5	5										*			* 2	* 2	* 3	* 4	0 4	0 5	1 7	2 9						
	Second	5	10													* 2	0 3	0 3	1 5	1 6	3 8	4 10	7 14						
	Third	5	15													0 2	0 3	1 4	2 6	3 8	6 10	8 13	13 19						
	Fourth	5	20													0 3	1 4	2 5	3 7	5 10	8 13	12 17	19 25						
	Fifth	5	25													1 3	2 4	3 6	4 8	7 11	11 15	17 20	25 29						
	Sixth	5	30													1 3	3 5	4 6	5 9	10 12	14 17	21 21	31 33						
	Seventh	5	35													2 3	4 5	6 7	6 10	13 14	18 19	25 26	37 38						
G	First	8	8									*			* 2	* 2	* 3	* 4	0 4	0 5	1 7	2 9							
	Second	8	16												* 2	0 3	0 3	1 5	1 6	3 8	4 10	7 14							
	Third	8	24												0 2	0 3	1 4	2 6	3 8	6 10	8 13	13 19							
	Fourth	8	32												0 3	1 4	2 5	3 7	5 10	8 13	12 17	19 25							
	Fifth	8	40												1 3	2 4	3 6	4 8	7 11	11 15	17 20	25 29							
	Sixth	8	48												1 3	3 5	4 6	5 9	10 12	14 17	21 21	31 33							
	Seventh	8	56												2 3	4 5	6 7	6 10	13 14	18 19	25 26	37 38							
H	First	13	13								*			* 2	* 2	* 3	* 4	0 4	0 5	1 7	2 9								
	Second	13	26											* 2	0 3	0 3	1 5	1 6	3 8	4 10	7 14								
	Third	13	39											0 2	0 3	1 4	2 6	3 8	6 10	8 13	13 19								
	Fourth	13	52											0 3	1 4	2 5	3 7	5 10	8 13	12 17	19 25								
	Fifth	13	65											1 3	2 4	3 6	4 8	7 11	11 15	17 20	25 29								
	Sixth	13	78											1 3	3 5	4 6	5 9	10 12	14 17	21 21	31 33								
	Seventh	13	91											2 3	4 5	6 7	6 10	13 14	18 19	25 26	37 38								
J	First	20	20							*			* 2	* 2	* 3	* 4	0 4	0 5	1 7	2 9									
	Second	20	40										* 2	0 3	0 3	1 5	1 6	3 8	4 10	7 14									
	Third	20	60										0 2	0 3	1 4	2 6	3 8	6 10	8 13	13 19									
	Fourth	20	80										0 3	1 4	2 5	3 7	5 10	8 13	12 17	19 25									
	Fifth	20	100										1 3	2 4	3 6	4 8	7 11	11 15	17 20	25 29									
	Sixth	20	120										1 3	3 5	4 6	5 9	10 12	14 17	21 21	31 33									
	Seventh	20	140										2 3	4 5	6 7	6 10	13 14	18 19	25 26	37 38									

Table 4.2(c)

Sample size code letter	Number of samples	Sample size	Accumulated sample size	0.010	0.015	0.025	0.040	0.065	0.10	0.15	0.25	0.40	0.65	1.0	1.5	2.5	4.0	6.5	10	15	25	40	65	100	150	250	400	650	1000
				A R	A R	A R	A R	A R	A R	A R	A R	A R	A R	A R	A R	A R	A R	A R	A R	A R	A R	A R	A R	A R	A R	A R	A R	A R	A R
K	First	32	32						·			· 2	· 2	· 3	· 4	0 4	0 5	1 7	2 9										
	Second	32	64									· 2	0 3	0 3	1 5	1 6	3 8	4 10	7 14										
	Third	32	96									0 2	0 3	1 4	2 6	3 8	6 10	8 13	13 19										
	Fourth	32	128									0 3	1 4	2 5	3 7	5 10	8 13	12 17	19 25										
	Fifth	32	160									1 3	2 4	3 6	5 8	7 11	11 15	17 20	25 29										
	Sixth	32	192									1 3	3 5	4 6	7 9	10 12	14 17	21 23	31 33										
	Seventh	32	224									2 3	4 5	6 7	9 10	13 14	18 19	25 26	37 38										
L	First	50	50					·			· 2	· 2	· 3	· 4	0 4	0 5	1 7	2 9											
	Second	50	100								· 2	0 3	0 3	1 5	1 6	3 8	4 10	7 14											
	Third	50	150								0 2	0 3	1 4	2 6	3 8	6 10	8 13	13 19											
	Fourth	50	200								0 3	1 4	2 5	3 7	5 10	8 13	12 17	19 25											
	Fifth	50	250								1 3	2 4	3 6	5 8	7 11	11 15	17 20	25 29											
	Sixth	50	300								1 3	3 5	4 6	7 9	10 12	14 17	21 23	31 33											
	Seventh	50	350								2 3	4 5	6 7	9 10	13 14	18 19	25 26	37 38											
M	First	80	80				·			· 2	· 2	· 3	· 4	0 4	0 5	1 7	2 9												
	Second	80	160							· 2	0 3	0 3	1 5	1 6	3 8	4 10	7 14												
	Third	80	240							0 2	0 3	1 4	2 6	3 8	6 10	8 13	13 19												
	Fourth	80	320							0 3	1 4	2 5	3 7	5 10	8 13	12 17	19 25												
	Fifth	80	400							1 3	2 4	3 6	5 8	7 11	11 15	17 20	25 29												
	Sixth	80	480							1 3	3 5	4 6	7 9	10 12	14 17	21 23	31 33												
	Seventh	80	560							2 3	4 5	6 7	9 10	13 14	18 19	25 26	37 38												
N	First	125	125			·			· 2	· 2	· 3	· 4	0 4	0 5	1 7	2 9													
	Second	125	250						· 2	0 3	0 3	1 5	1 6	3 8	4 10	7 14													
	Third	125	375						0 2	0 3	1 4	2 6	3 8	6 10	8 13	13 19													
	Fourth	125	500						0 3	1 4	2 5	3 7	5 10	8 13	12 17	19 25													
	Fifth	125	625						1 3	2 4	3 6	5 8	7 11	11 15	17 20	25 29													
	Sixth	125	750						1 3	3 5	4 6	7 9	10 12	14 17	21 23	31 33													
	Seventh	125	875						2 3	4 5	6 7	9 10	13 14	18 19	25 26	37 38													
P	First	200	200		·			· 2	· 2	· 3	· 4	0 4	0 5	1 7	2 9														
	Second	200	400					· 2	0 3	0 3	1 5	1 6	3 8	4 10	7 14														
	Third	200	600					0 2	0 3	1 4	2 6	3 8	6 10	8 13	13 19														
	Fourth	200	800					0 3	1 4	2 5	3 7	5 10	8 13	12 17	19 25														
	Fifth	200	1000					1 3	2 4	3 6	5 8	7 11	11 15	17 20	25 29														
	Sixth	200	1200					1 3	3 5	4 6	7 9	10 12	14 17	21 23	31 33														
	Seventh	200	1400					2 3	4 5	6 7	9 10	13 14	18 19	25 26	37 38														
Q	First	315	315	·			· 2	· 2	· 3	· 4	0 4	0 5	1 7	2 9															
	Second	315	630				· 2	0 3	0 3	1 5	1 6	3 8	4 10	7 14															
	Third	315	945				0 2	0 3	1 4	2 6	3 8	6 10	8 13	13 19															
	Fourth	315	1260				0 3	1 4	2 5	3 7	5 10	8 13	12 17	19 25															
	Fifth	315	1575				1 3	2 4	3 6	5 8	7 11	11 15	17 20	25 29															
	Sixth	315	1890				1 3	3 5	4 6	7 9	10 12	14 17	21 23	31 33															
	Seventh	315	2205				2 3	4 5	6 7	9 10	13 14	18 19	25 26	37 38															
R	First	500	500			· 2	· 2	· 3	· 4	0 4	0 5	1 7	2 9																
	Second	500	1000			· 2	0 3	0 3	1 5	1 6	3 8	4 10	7 14																
	Third	500	1500			0 2	0 3	1 4	2 6	3 8	6 10	8 13	13 19																
	Fourth	500	2000			0 3	1 4	2 5	3 7	5 10	8 13	12 17	19 25																
	Fifth	500	2500			1 3	2 4	3 6	5 8	7 11	11 15	17 20	25 29																
	Sixth	500	3000			1 3	3 5	4 6	7 9	10 12	14 17	21 23	31 33																
	Seventh	500	3500			2 3	4 5	6 7	9 10	13 14	18 19	25 26	37 38																

Acceptable Quality Level (normal control)

(i) *To the supplier*
A 'good' batch is rejected on the evidence of the sample, i.e. $k \geqslant R$ although $p \leqslant p_1$. This is called in statistics an error of the first kind, and the risk is expressed as a probability α defined by

$$\alpha = \text{prob}(k \geqslant R | p \leqslant p_1)$$

or, which is the same

$$1 - \alpha = \text{prob}(k \leqslant A | p \leqslant p_1)$$

(ii) *To the customer*
A 'bad' batch is accepted on the evidence of the sample, i.e. $k \leqslant R$ although $p \geqslant p_2$. This is a statistical error of the second kind, and the risk β is defined as

$$\beta = \text{prob}(k \leqslant R | p \geqslant p_2)$$

In general the aim is to minimize both risks in order to make the control process as fair as possible to both the supplier and the customer. The curve of Fig. 4.1 gives the probability p_a of accepting a batch as a function of the true fraction of defectives p, with these risks, corresponding to $p = p_1$ and $p = p_2$ respectively, shown.

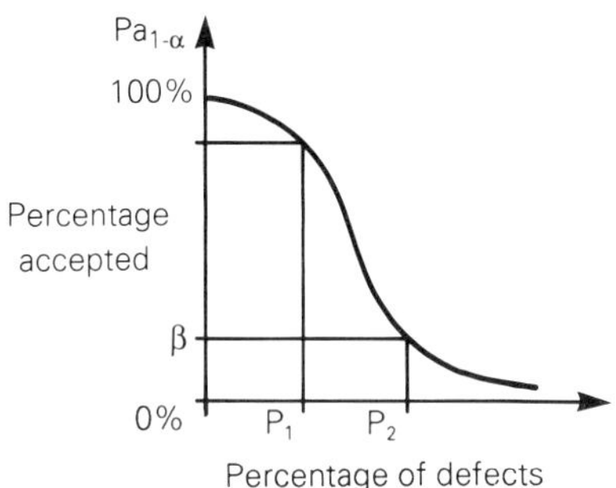

Figure 4.1 Effect of fraction of defectives on the probability of accepting a batch.

p_1 (or the percentage $P_1 = 100p_1$) is effectively equal to the acceptable quality level AQL and p_2 (or $P_2 = 100p_2$) to the tolerable quality level TQL, also called the limiting quality LQ. α measures the risk to the supplier, β that to the customer. Thus the pairs (α, P_1), (β, P_2) can be used as a basis for the control scheme.

4.1.4 Changing from one level of control to another

Assuming that the process starts in normal level, the rules are:

if normal and	10 successive batches are accepted change to relaxed
	2 successive batches are rejected change to strict
if relaxed and	1 batch is rejected change to normal
if strict and	5 successive batches are accepted change to normal

This is shown diagramatically in Fig. 4.2.

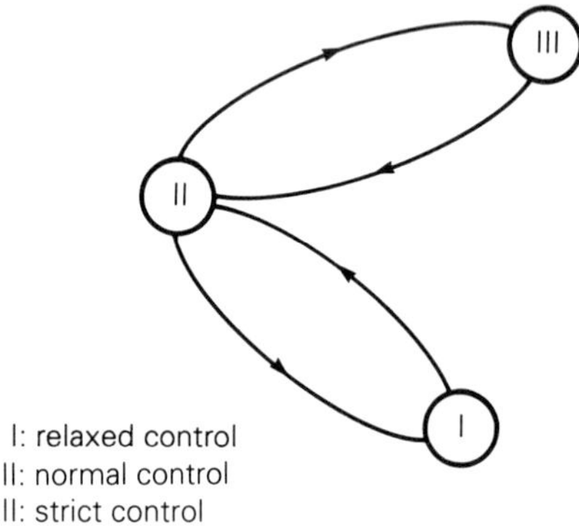

Figure 4.2 State change diagram for control levels.

4.1.5 Efficiency curves

A control procedure can be regarded as a filter that allows only those batches having a percentage of defectives not exceeding a certain value P_0 to pass through. A perfect, or 100% efficient, procedure would have the characteristics of Fig. 4.3, passing all batches having percentage defective $P \leqslant P_0$ and rejecting all with $P > P_0$. As we have seen, this is not attainable if the procedure is based on sampling; the curve of the fraction p_a accepted (or percentage $P_a = 100p_a$), based on one or other of the standard statistical laws, has the general form of Fig. 4.4. The closer this curve approximates to the rectangular form of Fig. 4.3 the more efficient is the procedure; but in general this increase in efficiency is gained at the cost of increasing the sample size n.

The statistical laws used in determining the curve are as follows:

- hypergeometric when the batch N is small and sampling is without replacement;
- binomial when $n/N < 0.1$, where n is the sample size;
- Poisson when the batch size is large but unknown.

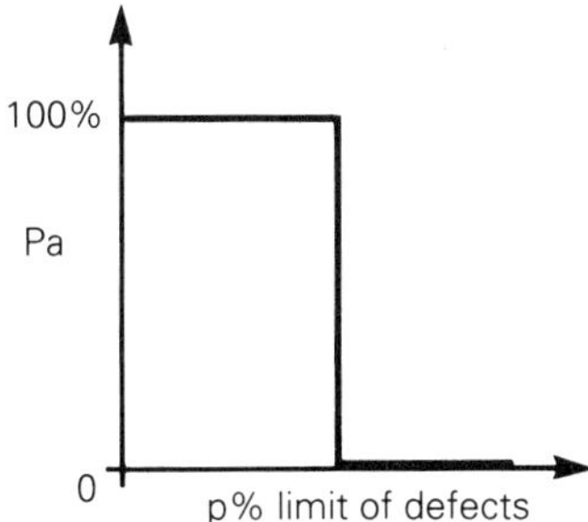

Figure 4.3 Idealized control system with 100% efficiency.

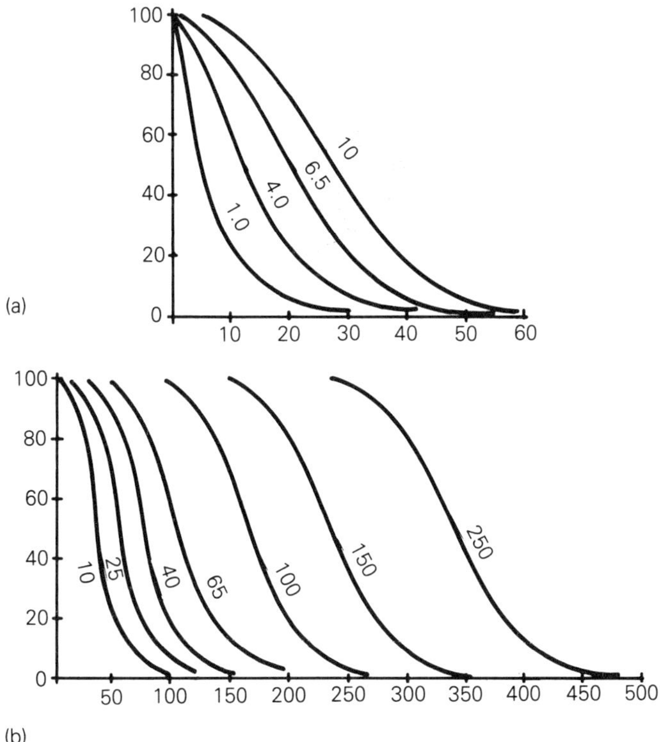

Figure 4.4 Efficiency curves for simple sampling plans (letter code E in Table 4.2a).

The probability p_a of accepting a batch containing a fraction p of defectives is the probability that the number k of defectives found in a random sample of n items does not exceed a stated number A; this is expressed formally as

$$p_a = \text{prob}(k \leqslant A|n, p)$$

Let p_1 be the greatest fraction of defectives that is acceptable; then p_a is given in terms of A and p_1 by the following:

1. For $n/N > 0.1$: hypergeometric law

$$p_a = \sum_{k=0}^{k=A} \binom{Np_1}{k} \binom{N(1-p_1)}{n-k} \Big/ \binom{N}{n}$$

2. For $n/N \leqslant 0.1$: binomial law

$$p_a = \sum_{k=0}^{k=A} \binom{n}{k} p_1^k(1 - p_1)^{n-k}$$

3. For N large, $p < 0.1$: Poisson law

$$p_a = \sum_{k=0}^{k=A} \frac{\exp(-m)m^k}{k!}$$

where if n is the sample size, $m = np$.

The efficiency curve can also be found experimentally, by testing numbers of batches with different but known fractions of defectives. Then

$$p_a = f(p) = \frac{\text{no. of batches accepted}}{\text{no. of batches tested}}$$

for each value of p. Efficiency curves relevant to current practice are given in the Standards documents; Fig. 4.4 and Table 4.3 are examples.

Table 4.3 Levels of acceptable quality

	Level of acceptable quality (normal control)			
P_a	1.0	4.0	6.5	10
	p (percent defective)			
99.0	0.077	1.19	3.63	7.0
95.0	0.394	2.81	6.83	11.3
90.0	0.807	4.16	8.80	14.2
75.0	2.19	7.41	13.4	19.9
50.0	5.19	12.6	20.0	27.5
25.0	10.1	19.4	28.0	36.2
10.0	16.2	26.8	36.0	44.4
5.0	20.6	31.6	41.0	49.5
1.0	29.8	41.5	50.6	58.7
	1.5	6.5	10.0	
	Level of acceptable quality (reinforced control)			

Example 1
Batch size $= 80$, AQL $= 1\%$.
From Table 4.1 the code letter for this batch size and a normal level of control is E (column II); then from Table 4.2(a) the sample size is 13 and the criteria are

$$\text{(accept) } A = 0 \qquad \text{(reject) } R = 1$$

This specifies the control procedure; the efficiency curve for 1% AQL is given in Fig. 4.4(a).

4.1.6 Quality improvement resulting from control

The control procedure can be operated in such a way that the stream of outgoing items (i.e. after the procedure has been applied) has a higher quality than the incoming stream. That is, if p and p' are the fractions of defectives before and after control, then $p' < p$. This can be achieved by testing *all* the rejected items and adding those found to be free of defects to the output stream.

Let p_a be the fraction of batches that are accepted by the control procedure: as we saw in section 4.1.5 this is a function of the (unknown) fraction p of defective items in the batch. Thus if we test r batches of N items each we accept $p_a r$ batches and reject $(1 - p_a)r$. The output stream at this stage consists of $p_a rN$ items among which are $pp_a rN$ defectives.

The reject stream consists of $(1 - p_a)rN$ items, of which $p(1 - p_a)rN$ are defective and $(1 - p)(1 - p_a)rN$ are free of defects. If we remove all the defectives and add the defect-free items to the 'accept' stream we have a total outgoing stream of $p_a rN + (1 - p)(1 - p_a)rN = (1 - p + pp_a)rN$ items, among which are pp_a defectives.

Thus the outgoing quality, the fraction p' of defectives in the output stream, is

$$p' = \frac{pp_a}{1 - p + pp_a} < p$$

since $1 - p > 0$. If, as expected, p is small and p_a is close to unity, p' is close to pp_a.

Since p_a is a function of p and can be calculated for a range of values of p when the details of the procedure are known (see section 4.1.5), p' also can be calculated as a function of p. The expression shows that $p' = 0$ when $p = 0$, which is obviously true (all the items are good), and $p' = p_a$ when $p = 1$; but then $p_a = 0$ (all items are bad and no batches are accepted), so $p' = 0$ again: not a meaningful result, since there are now no items in the output stream. It follows, however, that p' will have a maximum value for some value of p between 0 and 1 (i.e. of the percentage P between 0 and 100), which corresponds to a *minimum* outgoing quality after the test.

Figure 4.5 gives the curve of p' as a function of p for a control procedure of simple sampling with sample size $n = 20$ and acceptance level $A = 3$ defectives.

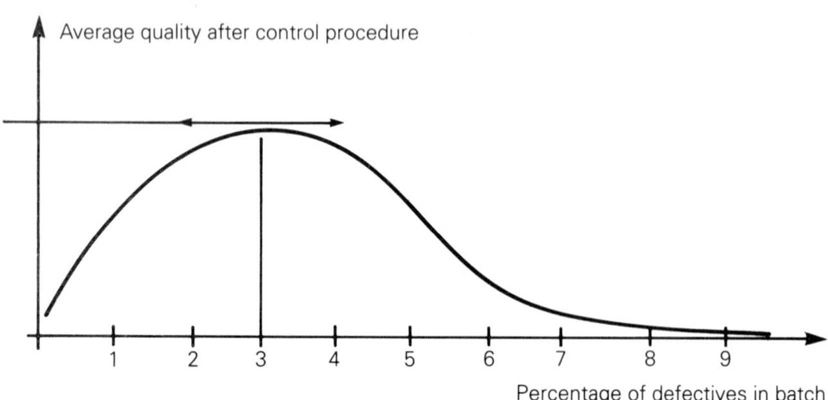

Figure 4.5 Effect of control: outgoing quality.

4.1.7 Average number of items tested

The average total number of items tested can be computed as a function of the true quality of the batches, for simple, double and multiple sampling procedures. This enables an economic choice of procedure to be made.

For simple sampling the size is n, the value chosen.

For *double sampling* let p_2' be the probability of needing to take a second sample.

$$p_2' = \text{prob}(A_1 < k_1 < R_1) = \sum_{kA_1}^{k=R_1} \binom{n}{k} p_k(1 - p_k)^{n-k}$$

where p is the true fraction of defectives in the batch. Then if n_1 and n_2 are the sizes of the first and second samples respectively the average number tested in this procedure is

$$n_{\text{av}} = n_1 + p_2' n_2$$

For *multiple sampling* with an obvious extension of the notation the result is

$$n_{av} = n_1 + p'_2 n_2 + p'_3 n_3 + \ldots + p'_8 n_8$$

If the procedure includes sorting the rejects to remove the defective items the average numbers handled become

- for simple sampling

$$n'_{av} = p_a n + p(1 - p_a)n$$

  where p_a is the probability of acceptance;
- for double sampling

$$n'_{av} = p_{a1} n_1 + p_{a2}(n_1 + n_2) + (1 - p_{a1} - p_{a2})N$$

  where p_{a1}, p_{a2} are the probabilities of acceptance for the samples n_1, n_2 respectively, and N is the number of items checked.

4.2 SEQUENTIAL TESTING: WALD'S TEST

Wald's test was developed with the aim of reducing the number of items tested; the principle is that items are drawn and tested one after another and the decision to accept or reject the batch is taken when one or other of two conditions is satisfied.

The procedure is shown graphically in Fig. 4.6. Items are tested in the order 1, 2, 3, . . . and the accumulated number of defectives after each test is plotted. This gives a stepped line, moving always either to the right or upwards, and the decision is taken when the line meets one or other of the 'accept' or 'reject' boundaries.

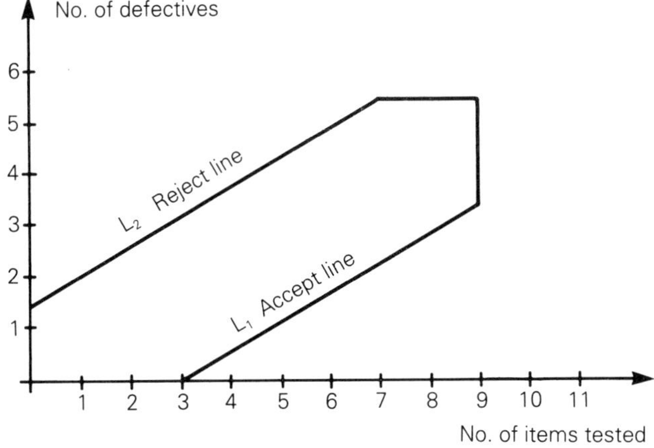

Figure 4.6 Limits for sequential sampling (Wald test).

4.2.1 Determining the boundaries

The boundaries are determined by the two risks α and β to the supplier and customer respectively (cf. section 4.1.3) and the corresponding probabilities p_1, p_2. The equations of the lines are as follows:

$$\text{(accept) L1} \qquad a_n = h_1 + Sn$$

$$\text{(reject) L2} \qquad r_n = h_2 + Sn$$

where

$$S = K \log\left(\frac{1 - p_1}{1 - p_2}\right)$$

$$h_1 = K \log\left(\frac{1 - \alpha}{\beta}\right)$$

$$h_2 = K \log\left(\frac{1 - \beta}{\alpha}\right)$$

$$K = \left[\log\left(\frac{p_2}{p_1}\right) + \log\left(\frac{1 - p_1}{1 - p_2}\right)\right]^{-1}$$

and logarithms are to base 10.

4.2.2 Efficiency

An estimate of efficiency is given by the curve shown earlier as Fig. 4.1.

4.2.3 Scoring procedure for Wald's test

Instead of the graphical method just described a scoring procedure can be used, as follows. As before, items are tested one after another.

We define a score H and two critical values H_1, H_2:

$$H_1 = \frac{h_1 + h_2}{S} \qquad H_2 = \frac{h_2}{S}$$

with h_1, h_2 and S as before. Initially $H = H_2$ (the 'handicap'); after each successive test

add 1 to the current value if the item is 'good'

subtract $(1 - S)/S$ if the item is 'bad'

and accept when $H = H_1$ or reject when $H = 0$.

4.3 CONTROL BY MEASURED PROPERTIES

For control by measured properties, obviously the property chosen as a criterion must be measurable; in general this method needs a smaller sample size than testing by attributes, for the same efficiency.

We make the basic assumption that the values of the measurement chosen are distributed normally (Gaussian) with mean m and standard deviation σ; we assume also that the true values of m and σ for the batch are not known and have therefore to be estimated from the sample.

With a sample of values $(x_1, x_2, \ldots, x_n)$ the estimates for m and σ are

$$\hat{m} = \frac{\Sigma x_i}{n}$$

$$\hat{\sigma} = \left(\frac{\Sigma(x_i - \hat{m})^2}{n - 1}\right)^{1/2}$$

Let T_u be the upper tolerance, i.e. the maximum acceptable value for the measure. Then the fraction p of defective items in the batch is given by the probability that a value exceeds T_u, i.e.

$$p = \text{prob}\left(U > \frac{T_u - \hat{m}}{\hat{\sigma}}\right)$$

where, as usual, U is the reduced normal variate $(x - m)/\sigma$ (Fig. 4.7). p is related directly to the AQL and is easily found from the table for the Gaussian distribution.

The calculations involved here are the same as for the control charts based on measured properties (cf. Chapter 3, section 3.2.1).

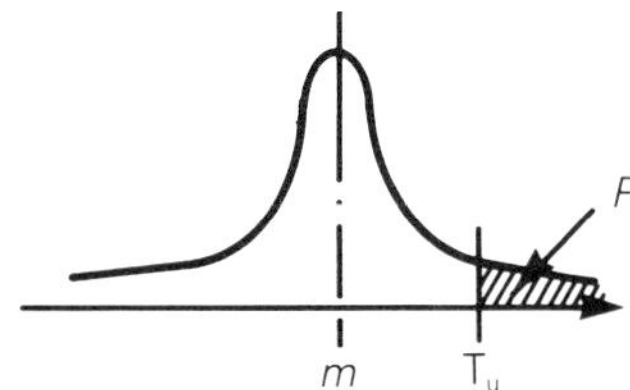

Figure 4.7 Upper tolerance level (Gaussian distribution).

4.4 SAMPLING PROCEDURES

There are many different ways in which samples can be taken, and the appropriate one to use in any particular case will depend on how the batch has been assembled: the sample should always reflect the reality of the situation as closely as possible.

In *random sampling* items are chosen with the help of random numbers, which can be obtained from tables (Appendix 8 gives a short table) or generated by a computer program – standard programs for this are available. This method is used when it can be assumed that the batch is homogeneous.

In *stratified sampling* the items in the batch come from several different sources ('strata') and it is desirable to take this into account in the sampling. This can be done by drawing random samples from each source and combining them to form the sample for the test. Examples of this situation are batches consisting of items made in different factories or in different runs in the same factory.

Two-level sampling consists in first selecting large samples from the primary units in the batch and then constituting the sample at random.

5

Cause-and-effect analysis

5.1 THE ISHIKAWA CAUSE–EFFECT DIAGRAM

5.1.1 General principles

When a manufacturing process is being monitored the first sign that something is wrong is the production of items that have to be scrapped; if further unwelcome effects are to be avoided the real causes of this must be discovered.

Since manufacturing processes often use complex systems, as many people as possible who are able to contribute to solving the problem should participate in the investigation, and in particular the users of the system. When the group has got together, notice should be taken of all suggestions concerning the loss of quality:

- variability of the raw materials;
- variability of the machines involved;
- changes in the workforce;
- changes in the working environment – e.g. from day shift to night shift;
- changes in working practices.

5.1.2 Application of these to the problem

Experience has shown that the causes of any effect in a manufacturing enterprise can be grouped into five main classes, which can be represented as a basic cause-and-effect diagram (Fig. 5.1). In the investigation each possible cause is recorded on this diagram. The next step is to establish the validity of the assertions and the relative importance of the various possible causes. They cannot all be investigated at once and so they must be put in order, and for this a scheme of weighted voting is helpful: each participant gives a weight to each cause and the causes that receive the greatest total weights are studied first.

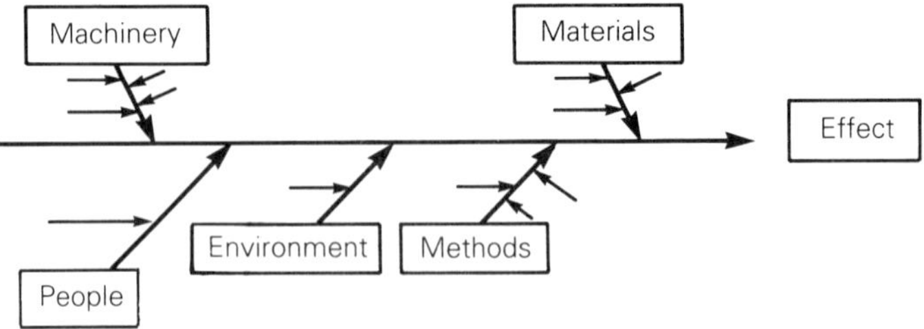

Figure 5.1 Typical cause/effect diagram for manufacture.

Example 1
The case studied is chosen to show how very general this approach is to the problems of industry. The problem is

'the coffee machine is giving bad coffee'

The possible causes listed are as follows:

- water is cold;
- water is too hot;
- water is polluted;
- water is chlorinated;
- poor brand of coffee;
- wrong amount of coffee;
- coffee badly ground;
- poor quality sugar (cane, white);
- too much sugar;
- oil on beaker;
- plastic beaker;
- water temperature;
- wrong amounts/proportions of ingredients;
- machine needs attention;
- machine inconvenient/unattractive;
- location of machine wrong – noise, dust etc.

Entering these on the basic diagram of Fig. 5.1 we get Fig. 5.2.

There were five participants in the study; each weighted each possible cause on a scale of 0–20 and the highest total scores were as follows:

- water quality (hard or soft) (80)
- brand and quantity of coffee (65)
- beaker (50)

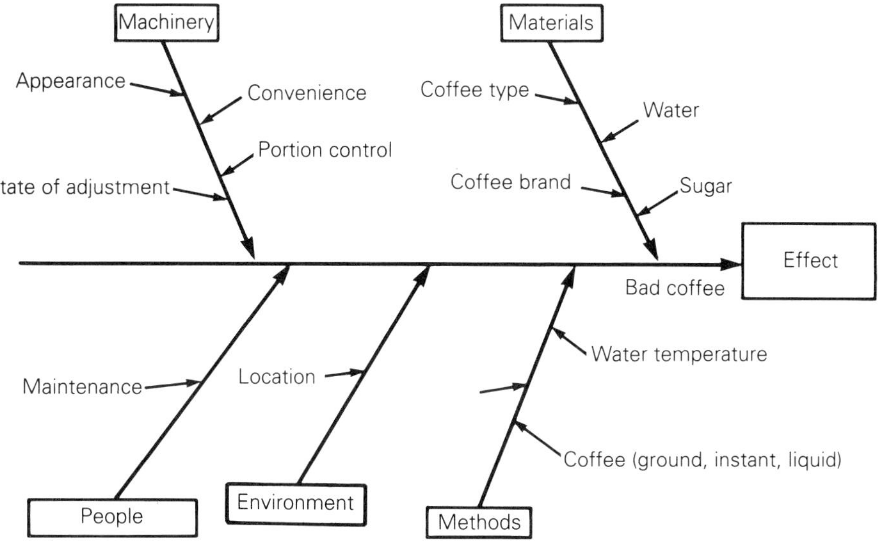

Figure 5.2 Cause/effect diagram for the coffee machine problem.

Having agreed on a short list of possible causes the usual next course is to make tests and experiments and carry out statistical analyses such as correlations, analysis of variance etc.

5.2 PARETO OR ABC ANALYSIS

Pareto or ABC analysis is an investigatory tool that enables the quality assurance service to assign priorities to the possible sources of quality defects — examination of rejects, for example, is the most expensive action. It can also be used to assess the improvement that has been achieved in any process by comparing the ABC curves for different dates. We describe it here in the context of quality in manufacturing production, and thus of rejects, but it is of much more general application.

5.2.1 The method

Rejects are put into classes according to some criterion and the classes are arranged in decreasing order of the costs they incur. This is then represented graphically – the ABC graph – with accumulated percentage of costs plotted against the accumulated percentage of types. Figure 5.3 shows the typical form of this curve.

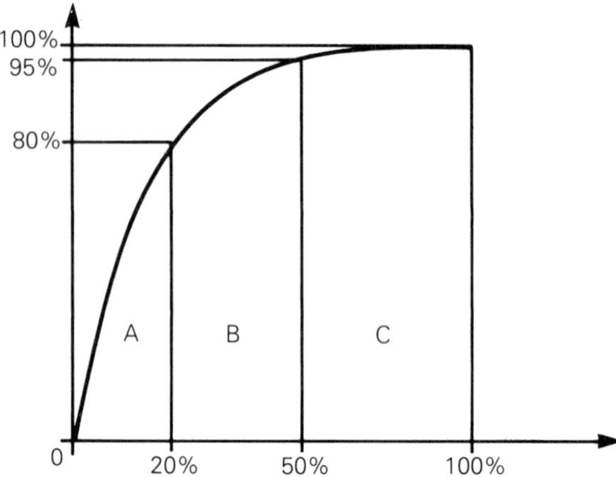

Figure 5.3 Pareto (ABC) analysis.

Zone A is the top priority zone. In most cases it is found that about 20% of the rejects account for about 80% of the costs.

Zone B contains the next 30% of rejects, which account for about 15% of the costs.

Zone C: the remaining 50% of rejects account for the remaining 5% of the costs.

Example 2

The costs associated with the rejects produced by the various machines in a manufacturing unit were as shown in Table 5.1. Putting these in order of cost we obtain Table 5.2 which, plotted, gives the curve of Fig. 5.4. The conclusion is that machines 11, 10, 1, 8, 9 and 3 should be examined first; restoring these to proper adjustment would save nearly 80% of the costs due to rejects.

Table 5.1 Costs per machine

Machine no.	Costs (£100)	Machine no.	Costs (£100)
1	100	8	80
2	32	9	55
3	50	10	150
4	19	11	160
5	4	12	5
6	30	13	10
7	40	14	20

Table 5.2 Data for the production of the curve in figure 5.4

Machine	Costs	Accumulated	Percentage
11	160	160	21.2
10	150	310	41.0
1	100	410	54.3
8	80	490	64.9
9	55	545	72.2
3	50	595	78.8
7	40	635	84.0
2	32	667	88.0
6	30	697	92.0
14	20	717	95.0
4	19	736	97.5
13	10	746	98.8
12	5	751	99.5
5	4	755	100.0

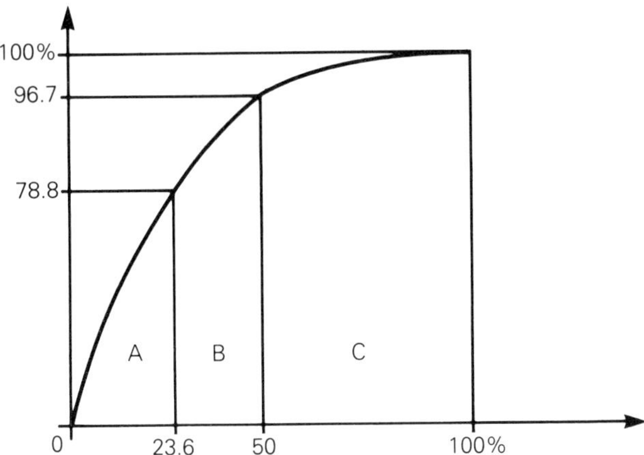

Figure 5.4 ABC analysis for Example 2.

Experience has shown that in general the ABC curve will have one or other of the three forms of Fig. 5.5:

Form 1 the division into classes is very sharp, with Zone A dominating the costs.
Form 2 less sharp division
Form 3 no order of priority

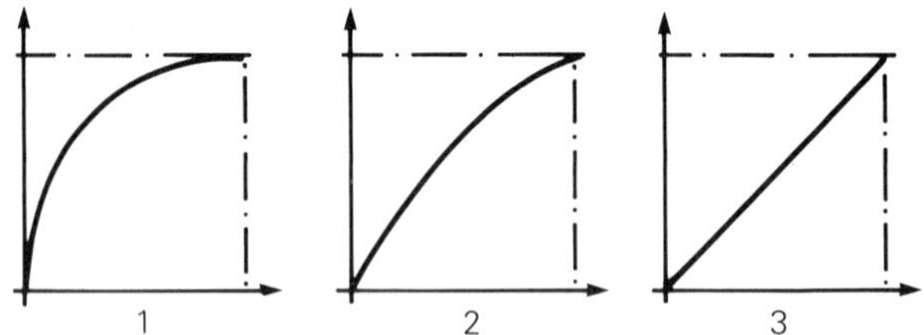

Figure 5.5 The three forms of the ABC curve.

5.3 RANK CORRELATION: SPEARMAN'S COEFFICIENT ρ_S

Spearman's coefficient is a measure of the relation between pairs of quantities that makes no assumption about their being normally distributed and can be used with both continuous and discrete variables. All that is necessary is that the values of the variables in question can be put in order, either increasing or decreasing: this is the reason for the term 'rank'.

5.3.1 The method

Let X, Y be the variables whose relationship we wish to investigate. The procedure is as follows.

1. The values of X and Y are put in increasing order; suppose there are n.
2. The first (smallest) value of X is given rank 1, the next largest rank 2 and so on until all have been ranked; if there are equal values each is given the average rank for the group. This is repeated for Y.
3. For each pair (X_i, Y_i) the difference d_i of the ranks is calculated.
4. ρ_S is calculated from

$$\rho_S = 1 - \frac{6\Sigma d_i^2}{n^3 - n}$$

(a) *Interpretation*
ρ_S has a value between -1 and $+1$; if

$\rho_S = 0$ there is no correlation between X and Y
$= 1$ X and Y are strongly correlated and increase or decrease together
$= -1$ X and Y are strongly correlated and increase or decrease in opposite directions

A non-zero value can arise by chance even when X, Y are completely uncorrelated, and so a test for the significance of the value found is needed. If there are more than 10 pairs (X_i, Y_i) Student's test is applicable, as follows.

The quantity

$$T = \frac{(n - 2)\rho^2}{(1 - \rho^2)^{1/2}}$$

has a Student distribution with $n - 2$ degrees of freedom; thus the hypothesis that ρS is zero (i.e. that the variables are not correlated) is rejected at level α if $T > t(n - 2, 1 - \alpha)$.

5.3.2 Program for computing ρ_S

A program for computing ρ_S is given in Fig. 5.6.

```
10  CLS
20  LOCATE 12,15:PRINT "RANK CORRELATION"
30  LOCATE 13,15:PRINT "________________________"
40  LOCATE 15,20:PRINT "SPEARMAN COEFFICIENT"
50  FOR E=1 TO 3000:NEXT E
60  PRINT:PRINT
70  CLS:LOCATE 6,5:INPUT "NO. OF OBSERVATIONS     ",N
80  LOCATE 9,5:INPUT "NAME OF FIRST VARIABLE     :",X$
90  PRINT
100 LOCATE 12,5:INPUT "NAME OF SECOND VARIABLE        :",V$
110 DIM R1(N),R2(N),S(N)
120 FOR I=1 TO N
130 CLS:LOCATE 4,5:PRINT "OBSERVATION      ";I;"   :"
140 LOCATE 6,5:PRINT "RANK ACCORDING TO CLASSIFICA
    TION   ";X$;TAB(40)
150 INPUT R1(I)
160 LOCATE 8,5:PRINT "RANK ACCORDING TO CLASSIFICA
    TION   ";V$;TAB(40)
170 INPUT R2(I)
180 PRINT
190 NEXT I
200 D2=0
210 FOR I=1 TO N
220 D2=D2+(R1(I)-R2(I))^2
230 NEXT I
240 R1=1-(6*D2)/(N^3-N)
250 PRINT:PRINT
260 PRINT "OBSERVATIONS   ",X$,Y$
270 PRINT
```

Figure 5.6 BASIC program for the calculation of the Spearman correlation.

```
280 FOR I=1 TO N
290 PRINT TAB(2);I,R1(I),R2(I)
300 NEXT I
310 PRINT:PRINT
320 PRINT "SPEARMAN COEFFICIENT=    ";
    INT(10000*R1+.5)/100;"%"
330 END
```

Figure 5.6 (continued)

5.4 ANALYSIS OF VARIANCE

Correlation analysis takes into account only one factor at a time and therefore cannot measure the interaction of two factors. For this the method of analysis of variance is used, which enables us

- to study the simultaneous effects of several factors
- to reveal any interactions between different factors
- to optimize the number of observations needed

It thus seems particularly well adapted to investigating the reasons why reject items are being produced.

5.4.1 Mathematics of the method

Analysis of variance is based on these assumptions:

- the factors affect only the means of the observations and not their variances;
- the effects of the different factors are additive;
- the residual variations (the 'errors') in the observations, after the effects of the factors have been taken into account, are distributed normally with zero mean.

(a) *The general linear model*
We consider the case of a quantity Y that is affected by two factors A and B which can interact; is Y_{ij} is a value observed for Y when A and B have values A_i and B_j respectively, the model is

$$Y_{ij} = m + \alpha_i + \beta_j + \gamma_{ij} + e_{ij}$$

where α_i and β_j measure the effects of A, B respectively, γ_{ij} measures the effect of the interaction between A and B, e_{ij} is the residual error and m is a constant.

In order to study the interaction effect we must have repeated observations with the same values of A and B. The data can be set out as in Table 5.3 in which the following notation is used.

Table 5.3 Notation for two-factor analysis of variance

Factor A	Factor B					
	B_1	B_2	B_j	B_s	Total	Mean
A_1	y_{111} y_{11n}	y_{121} y_{12n}	y_{ij1} y_{1jn}	y_{1s1} y_{1sn}	$\overline{y_{1..}}$	$\overline{y_{1..}}$
A_2	y_{211} y_{21n}	y_{221} y_{22n}	y_{2j1} y_{2jn}	y_{2s1} y_{2sn}	$y_{2..}$	$\overline{y_{2..}}$
A_i	y_{i11} y_{i1n}	y_{i21} y_{i2n}	y_{ij1} y_{ijn}	y_{is1} y_{isn}	$y_{i..}$	$\overline{y_{i..}}$
A_r	y_{r11} y_{r1n}	y_{r21} y_{r2n}	y_{rj1} y_{rjn}	y_{rs1} y_{rsn}	$y_{r..}$	$\overline{y_{r..}}$
Total	$y_{.1.}$	$y_{.2.}$	$y_{.j.}$	$y_{.s.}$	$\overline{y_{...}}$	
Mean	$\overline{y_{.1.}}$	$\overline{y_{.2.}}$	$\overline{y_{.j.}}$	$\overline{y_{.s.}}$		$\overline{y_{...}}$

$A_1, A_2, \ldots A_r$ are the values of A
$B_1, B_2, \ldots B_s$ are the values of B

(These need not be numerical values. For example, they can label classes to which the factors belong.)

Y_{ijk} $(k = 1, 2, \ldots, n)$ is the result of the kth observation with $A = A_i$, $B = B_j$

$Y_{ij.} = \sum_k Y_{ijk}$ is the sum of the values corresponding to (A_i, B_j) i.e. in one cell of the table

$\bar{Y}_{ij.} = \sum_k \frac{Y_{ij}}{n}$ is the mean of these values

$Y_{i..} = \sum_{j,k} Y_{ijk}$ is the sum of the values corresponding to A_i, i.e. the row sum

$\bar{Y}_{i..} = \sum_{j,k} \frac{Y_{ijk}}{n}$ is the corresponding mean

$Y_{.j.} = \sum_{i,k} Y_{ijk}$ is the sum of the values corresponding to B_j, i.e. the column sum

$\bar{Y}_{.j.} = Y_{.j.}/nr$ is the corresponding mean

$Y_{...} = \sum_{i,j,k} Y_{ijk}$ is the sum of all the values

$\bar{Y}_{...} = Y_{...}/nrs$ is the overall mean

$Y_{ijk} - \bar{Y}_{...}$ is the deviation of an individual observation from the overall mean; the method considers the sum of the squares of all these deviations in order to attribute the total to the separate effects of A, B and the interaction AB respectively. It uses the identity

$$Y_{ijk} - \bar{Y}_{...} = (Y_{ijk} - \bar{Y}_{ij}) + [(\bar{Y}_{ij} - \bar{Y}_{i..}) - (\bar{Y}_{.j.} - \bar{Y}_{...})] + (\bar{Y}_{i..} - \bar{Y}_{...}) + (\bar{Y}_{.j.} - \bar{Y}_{...})$$

Squaring both sides and summing over i, j, k we get

$$\sum_{ijk} (Y_{ijk} - \bar{Y}_{...})^2 = ns\sum_{i} (\bar{Y}_{i..} - \bar{Y}_{...})^2 + nr\sum_{j} (\bar{Y}_{.j.} - \bar{Y}_{...})^2 + n\sum_{ij} [(\bar{Y}_{ij.} - \bar{Y}_{i..}) - (\bar{Y}_{.j.} - \bar{Y}_{...})]^2 + \sum_{ijk} (Y_{ijk} - \bar{Y}_{ij.})^2$$

since all the sums of cross-products vanish. We can write this as

$$\text{SST} = \text{SSA} + \text{SSB} + \text{SSAB} + \text{SSE}$$

where SST is the sum of the squares of all the deviations from the mean, SSA, SSB and SSAB are the contributions to SST that can be attributed to the effects of A, B and the interaction AB respectively and SSE is the residual (random) effect. This equation 'analyses' the total variance into the separate components; it can be shown that these components have a χ^2 distribution with the following degrees of freedom:

SSA	$\mu_A = r - 1$
SSB	$\mu_B = s - 1$
SSAB	$\mu_{AB} = (r - 1)(s - 1)$
SST	$\mu_{\mathrm{T}} = nrs - 1$
SSE	$\mu_{\mathrm{E}} = (n - 1)rs$

Dividing each of the sums of squares by its degrees of freedom gives an estimate of its variance; all this is summarized in Table 5.4.

If A, B and the interaction AB have no effect on the observations – the 'null hypothesis' – the values SA^2 etc. are all estimates of the residual variance. It can be shown that in this case the ratios SA^2/SE^2,

Table 5.4 Two-factor analysis of variance: basic formulae

Source of effect	*Sums of squares*	*Degrees of freedom*	*Estimate of variance of error*
Factor A	$\mathrm{SSA} = sn \sum_{i}^{r} (\bar{Y}_{i..} - \bar{Y}_{...})^2$	$r - 1$	$\mathrm{SA}^2 = \frac{\mathrm{SSA}}{r - 1}$
Factor B	$\mathrm{SSB} = rn \sum_{j}^{s} (\bar{Y}_{.j.} - \bar{Y}_{...})^2$	$s - 1$	$\mathrm{SB}^2 = \frac{\mathrm{SSB}}{s - 1}$
Interaction AB	$\mathrm{SSAB} = n \sum_{i}^{r} \sum_{j}^{s} (\bar{Y}_{ij.} - \bar{Y}_{i..} - \bar{Y}_{.j.} + \bar{Y}_{...})^2$	$(r - 1)(s - 1)$	$\mathrm{SAB}^2 = \frac{\mathrm{SSAB}}{(r - 1)(s - 1)}$
Residual error	$\mathrm{SSE} = \sum_{i}^{r} \sum_{j}^{s} \sum_{k}^{n} (Y_{ijk} - \bar{Y}_{ij.})^2$	$rs(n - 1)$	$\mathrm{SE}^2 = \frac{\mathrm{SSE}}{rs(n - 1)}$

SB^2/SE^2 and SAB^2/SE^2 all have a Fisher distribution with degrees of freedom (ν_A, ν_E), (ν_B, ν_E), (ν_{AB}, ν_E) respectively.

(b) *Significance test for the factors*
Given this last result and a table of the Fisher distribution – one is given in Appendix 4 – a significance test for the null hypothesis is easily constructed.

However, if there is an interaction effect the separate effects of the factors are not additive and one of the hypotheses on which the method is based is not valid; in such a case we cannot measure the effects of the separate factors. It is therefore usual to start by investigating the interaction.

(c) *Testing the null hypothesis*
The Fisher function is denoted by $F(\nu_1, \nu_2; \theta)$. The significance test for the interaction is: if

$$SAB^2/SE^2 > F[(r-1)(s-1), rs(n-1); 1-\alpha]$$

then the hypothesis that the interaction AB has no effect is rejected at probability level α. The conclusion is therefore that there may be an interaction effect.

Similarly, for the factor A, if

$$SA^2/SE^2 > F[(r-1), rs(n-1); 1-\alpha]$$

then the hypothesis that A has no effect is rejected at probability level α, and so A may have a significant effect. Similarly for B, if $SB^2/SE^2 > F[(s-1), rs(n-1); 1-\alpha]$.

5.4.2 Program for two-factor analysis of variance

A BASIC program is given in Fig. 5.7. When this is supplied with the experimental data as input it outputs the values of $F(A)$, $F(B)$, $F(AB)$ for application of the Fisher test.

Example 3
The engineer responsible for the maintenance of a group of production machines wishes to know if the monthly cost of the substandard items produced by a machine is related to its age and/or its total annual production.

Taking factor A to be the age of the machine and factor B its annual production the data are as in Table 5.5. The program gives these results:

$SAB^2/SE^2 = 2.9$ no influence at the level $\alpha = 0.025$
$SA^2/SE^2 = 8.02$ weak influence
$SB/SE^2 = 258.5$ strong influence

```
10  CLS
20  REM "ANALYSIS OF VARIANCE - 2 FACTORS (R AND C),
    DIFFERENT MODES, N REPETITIONS"
30  LOCATE 10,8:PRINT "ANALYSIS OF VARIANCE 2 FACTORS"
40  DIM Y(20,20,5)
50  Y000=0
60  YIJK2=0
70  LOCATE 12,7:INPUT "r (number of modes of factor A)";R
80  LOCATE 14,7:INPUT "c (number of modes of factor B)";C
90  LOCATE 16,7:INPUT "n (number of repetitions)";N
100 LOCATE 18,7:PRINT "Enter Y(i,j;k) one line at a time"
110 FOR I=1 TO R
120 FOR K=1 TO N
130 FOR J=1 TO C
140 INPUT "x";X
150 Y(I,J,K)=X
160 YIJK2=YIJK2+Y(I,J,K)^2
170 Y000=Y000+Y(I,J,K)
180 NEXT J
190 NEXT K
200 NEXT I
210 YI=0:YI002=0
220 FOR I=1 TO R
230 FOR J=1 TO C
240 FOR K=1 TO N
250 YI=YI+Y(I,J,K)
260 NEXT K
270 NEXT J
280 YI002=YI002+YI^2
290 YI=0
300 NEXT I
310 YIJ=0
320 YJ=0:YJ002=0:YIJ02=0
330 FOR J=1 TO C
340 FOR I=1 TO R
350 FOR K=1 TO N
360 YJ=YJ+Y(I,J,K)
370 YIJ=YIJ+Y(I,J,K)
380 NEXT K
390 YIJ02=YIJ02+YIJ^2
400 YIJ=0
410 NEXT I
420 YJ002=YJ002+YJ^2
430 YJ=0
440 NEXT J
450 SST=YIJK2-(Y000^2)/(N*R*C)
```

Figure 5.7 BASIC program for two-factor analysis of variance.

```
460 SSA=YI002/(C*N)-(Y000^2)/(N*R*C)
470 SSB=YJ002/(N*R)-(Y000^2)/(N*R*C)
480 SSE=YIJK2-YIJ02/N
490 SSAB=SST-SSA-SSB-SSE
500 PRINT "Calculation of variances",
510 SA2=SSA/(R-1)
520 SB2=SSB/(C-1)
530 SAB2=SSAB/((R-1)*(C-1))
540 SE2=SSE/(R*C*(N-1))
550 IF N=1 THEN GOTO 560 ELSE GOTO 590
560 SE2=SAB2
570 SAB2=0
580 PRINT "NB If there are no repetitions, SE is estimated by SAB and
    SABn doesn't exist"
590 PRINT "SA2";SA2
600 PRINT "SB2";SB2
610 PRINT "SAB2";SAB2
620 PRINT "SE2";SE2
630 FA=SA2/SE2
640 FB=SB2/SE2
650 FAB=SAB2/SE2
660 PRINT "FA";FA,"FB";FB,"FAB";FAB
670 END
```

Figure 5.7 (continued)

Table 5.5 Data for Example 3 (p. 110)

	*B*1: 20 000	*B*2: 50 000	B3: 80 000
*A*1: < 3 yr	$Y_{111} = 20$	$Y_{121} = 35$	$Y_{131} = 60$
	$Y_{112} = 23$	$Y_{122} = 32$	$Y_{132} = 58$
*A*2: 3–6 yr	$Y_{211} = 21$	$Y_{221} = 36$	$Y_{231} = 48$
	$Y_{212} = 19$	$Y_{222} = 32$	$Y_{232} = 55$
*A*3: > 6 yr	$Y_{311} = 28$	$Y_{321} = 40$	$Y_{331} = 70$
	$Y_{312} = 28$	$Y_{322} = 45$	$Y_{332} = 75$

5.5 EXPERIMENTAL DESIGNS OF TYPE 2^n

Designs of type 2^n are designs in which there are n factors, each of which can be present at either of two levels. These levels need not be specified quantitatively, the only requirement being the possibility of distinguishing between the two – for example, for a lubricating oil, between high and low viscosity; for a transaction-processing system, between high message rate (say 3000 transactions per minute) and low message rate (say 500 transactions per minute). The aims of this design are as follows:

- to minimize the number of tests needed;
- to quantify the effect of each factor;
- to quantify the residual variance, i.e. the 'error' observed after the effects of the known factors have been accounted for;
- to reveal any interactions between the factors.

5.5.1 Designs without replication

In a complete design, i.e. where every factor enters at both levels, there are 2^n observations. Thus for three factors there are eight observations. Reduced layouts can be designed that require only 2^{n-1} observations, but for these the estimation of the effects of the various factors is more difficult.

5.5.2 Designs with replication: the type 2^{n+r}

Here each experiment is repeated r times; this enables the effects of factors not taken into account explicitly to be measured, and so the residual variance can be estimated.

Notation

Upper case letters A, B, C etc. denote the various factors; the two levels at which each can be present are denoted by + (higher), − (lower).

In Yates's convention a lower case letter, e.g. a, indicates that that factor, e.g. A, is present at its upper level and that all the factors not mentioned are at their lower levels. This is shown in Table 5.6. It simplifies the specification of the various combinations.

Table 5.6 Yates' table of signs for 2^3 design

Symbols	Sign combinations for effects of factors and interactions						
	A	B	C	AB	AC	BC	ABC
(1)	−	−	−	+	+	+	−
a	+	−	−	−	−	+	+
b	−	+	−	−	+	−	+
ab	+	+	−	+	−	−	−
c	−	−	+	−	+	−	+
ac	+	−	+	−	+	−	−
bc	−	+	+	−	−	+	−
abc	+	+	+	+	+	+	+

5.5.3 The 2^3 design

(a) *The Yates table*
In Table 5.6 the row labelled (1) gives the results of experiments in which all the factors are at their lowest levels (−). In row a, A is at its high level (+), B and C are low (−) etc. The table simply gives the rules for the signs for constructing the expressions of section 5.5.3(b) for the effects of A, B etc.

(b) *Estimates of the effects of the factors*
An estimate of the effect of any factor is the difference between the results of the observations made with that factor at its highest and lowest levels. Thus for the above design,

$$\text{effect of } A = \tfrac{1}{4}[(a) + (ab) + (ac) + (abc) - (1) - (b) - (c) - (bc)]$$

and

$$\hat{\alpha} = \tfrac{1}{2}(\text{effect of } A)$$

$$\text{effect of } B = \tfrac{1}{4}[(b) + (ab) + (bc) + (abc) - (a) - (c) - (ac)]$$

$$\hat{\beta} = \tfrac{1}{2}(\text{effect of } B)$$

etc.

The effect of the interaction AB is obtained by the rule of signs in the table:

$$\text{effect of } AB = \tfrac{1}{4}[(1) - (a) - (b) + (ab) + (c) - (ac) - (bc) + (abc)]$$

$$\widehat{\alpha\beta} = \tfrac{1}{2}(\text{effect of } AB)$$

(c) *Treatment of the residual variance*
The residual variance represents the combined effects of all the factors that are unknown or otherwise not under control. If the investigation is to give meaningful results it is important that these residual influences are randomized as far as possible; otherwise there is a risk that some results will be unreliable and the residual variance will be increased.

(i) *Design with replications*
The estimate of the residual variance is

$$\hat{\sigma}^2 = \sum_{ijk} \frac{(d_{ijku})^2}{n - k - 1}$$

where $d_{ijku} = Y_{ijku} - \hat{Y}_{ijk}$ and $n-k-1$ is the number of degrees of freedom.

(ii) *Design without replications*
The estimator is the higher order interaction; thus for the 2^3 design

$$\hat{\sigma}^2 = \widehat{\alpha\beta\gamma}$$

(d) *Three-factor model*
The model is

$$Y_{ijk} = Y_{...} + \hat{\alpha}X_{1i} + \hat{\beta}_{2j} + \alpha\beta X_{1i}X_{2j} + \hat{\gamma}_{3k} + \alpha\gamma X_{1i}X_{3k} + \beta\gamma X_{2j}X_{3k} + d_{ijk}$$

where Y_{ijk} is the true value at the point i, j, k, $\hat{Y}_{ijk}$ is the estimated value at that point and d_{ijk} is the error term.

$$Y_{ijk} = \hat{Y}_{ijk} + d_{ijk}$$

d_{ijk} serves to estimate the variance about the regression line:

$$(\hat{\sigma}_\gamma)^2 = \frac{\Sigma(d_{ijk})^2}{1}$$

where the denominator is 1 because with eight results there is only one degree of freedom for the variance.

In fact, it is the second order interaction that provides an estimate for $(\sigma_\gamma)^2$:

$$(\hat{\sigma}_\gamma)^2 = 8(d_{ijk})^2 \text{ with } d_{ijk} = \pm\alpha\beta\gamma$$

i.e.

$$(\hat{\sigma}_\gamma)^2 = 8(\pm\alpha\beta\gamma)^2$$

(i) *Significance tests*
To test the significance of the coefficients $\hat{\alpha}$, $\hat{\beta}$, $\hat{\gamma}$ and $\alpha\beta$ we need estimates for their standard deviations $\hat{\sigma}_\alpha$, $\hat{\sigma}_\beta$ etc; in this case these are

$$\hat{\sigma}_\alpha = \hat{\sigma}_\beta = \hat{\sigma}_{\alpha\beta} = \frac{\hat{\sigma}_\gamma}{\sqrt{2^3}} = |d_{ijk}| = |\widehat{\alpha\beta\gamma}|$$

The test is based on the result that $\hat{\alpha}/\hat{\sigma}_\alpha$, $\hat{\beta}/\hat{\sigma}_\beta$, $\hat{\gamma}/\hat{\sigma}_\gamma$ and $\alpha\beta/\hat{\sigma}_{\alpha\beta}$ all have a Student distribution with one degree of freedom.

By similar processes we can construct experimental designs of type 2^n with $n = 2, 4, 5, 6, \ldots$.

Example 4
The following results were obtained in a 2^3 experimental design.

(1)	*a*	*b*	*ab*	*c*	*ac*	*bc*	*abc*
15	18	10	20	15	14	8	17

Table 5.7 is the application of the Yates table of signs to these values, with the sums corresponding to A, B, AB etc, together with the ratios of the estimated coefficients and standard deviations. The significances of these are tested by comparing them with the Student distribution with one degree of freedom. For a probability level 0.05 we find, from the table of Appendix 2, $t(0.05;\ 1) = 6.3$, and therefore we conclude that the only significant coefficients are those of A and AB.

Thus the model for these results is

$$\hat{Y}_{ijk} = 14.62 + 7.0X_{1i} + 5.7X_{1i}X_{2j}$$

Table 5.7 Yates' table for Example 4 (p. 115)

Combinations	Results	Total	A	B	C	AB	AC	BC	ABC
(1)	15	+	−	−	−	+	+	+	−
a	18	+	+	−	−	−	−	+	+
b	10	+	−	+	−	−	+	−	+
ab	20	+	+	+	−	+	+	−	−
c	15	+	−	−	+	+	−	−	+
ac	14	+	+	−	+	−	+	−	−
bc	8	+	−	+	+	−	−	+	−
abc	17	+	+	+	+	+	+	+	+
Σ		117	21	−7	−9	17	−5	−1	3
Total			5.2	−1.75	−2.25	4.25	1.25	0.25	0.75
Coef.		14.52	2.62	−0.87	−1.12	2.12	0.62	0.125	0.375

5.6 GRAPHICAL METHOD: SCATTER DIAGRAM

A plot of one variable against another gives a visual impression of the relation between them, and can give an idea of the degree of correlation. This is illustrated in Fig. 5.8.

Example 5

The following table gives the aptitude test scores for each of a number of employees and the production levels achieved by each over a certain period. The question is, is there any relation between these numbers?

Employee	1	2	3	4	5	6	7	8	9
Test score	17	8	5	19	15	7	14	13	12
Production	150	90	70	200	140	100	130	150	110

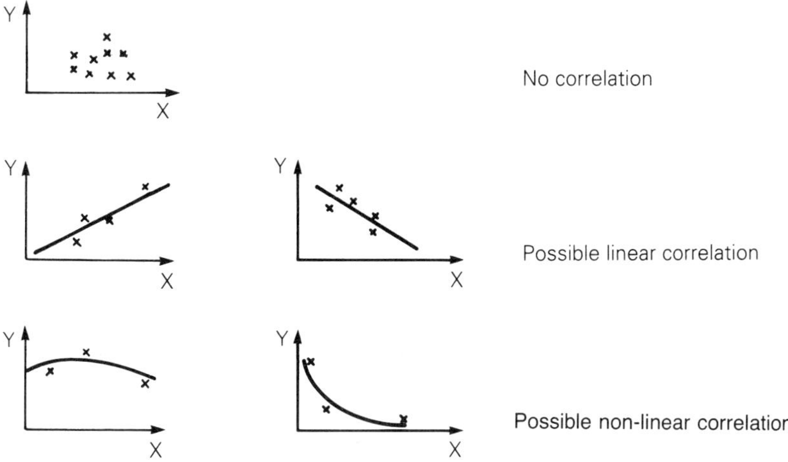

Figure 5.8 Examples of scatter diagrams.

The graph of Fig. 5.9 suggests that the two are related.

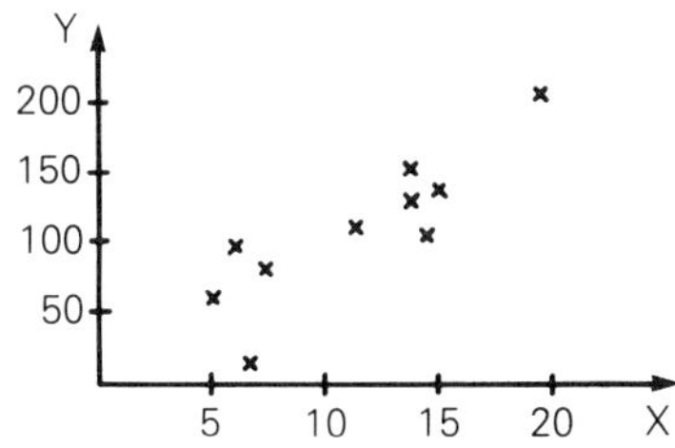

Figure 5.9 Scatter diagram for Example 3 (p. 116).

6

Basic mathematics

6.1 PROBABILITY: THEORY, DEFINITIONS

The probability $P(E)$ of an event E is a number lying between 0 and 1 that measures, in some sense, the likelihood of the event occurring. It can be derived by enumerating all the possibilities:

$$P(E) = \frac{\text{number of cases favourable to } E}{\text{total number of possible cases}}$$

This gives the 'true' probability of E; the difficulty is to carry out the enumeration. The alternative is an experimental procedure based on observations: if a number of observations are made, in each of which E might occur, and if $f_n(E)$ is the fraction of observations in which E is observed to occur, then

$$P(E) = \lim_{n\to\infty} f_n(E)$$

This is the 'frequency' definition of probability; the greater the number of observations, the closer the frequency approaches the 'true' probability.

As we said, $0 \leqslant P(E) \leqslant 1$; conventionally, $P(E) = 0$ corresponds to certainty that E will not occur, $P(E) = 1$ to certainty that it will occur.

(i) *Joint occurrence (intersection) of events*
The probability that both events A and B occur is

$$P(A \cap B) = P(A)P(B|A)$$

where $P(B|A)$ means the probability that B occurs, given that A occurs.

The general relation for n events $A_1, A_2, \ldots, A_n$ is

$$P(A_1 \cap A_2 \cap \ldots \cap A_n) = P(A_1)P(A_2|A_1)P(A_3|A_1A_2) \ldots P(A_n|A_1A_2 \ldots A_n)$$

If the events A, B are independent, meaning that the occurrence of either has no influence on the occurrence of the other, then

$$P(A \cap B) = P(A)P(B)$$

and in general, for n independent events,

$$P(A_1 \cap A_2 \cap \ldots \cap A_n) = \prod_{i=1}^{n} (A_i)$$

When the events under consideration are breakdowns of equipment the assumption of independence is usually valid.

(ii) *Occurrence of one or other of several possible events (union of events)*

$$P(A \cup B) = P(A) + P(B) - P(A \cap B)$$

(as can be seen by considering the number of possible occurrences of A and B).

If A, B are independent

$$P(A \cup B) = P(A) + P(B) - P(A)P(B)$$

and if A, B are incompatible (mutually exclusive)

$$P(A \cap B) = P(A) + P(B)$$

since then $P(A \cap B) = 0$. In general, for n events,

$$P(A_1 \cup A_2 \cup \ldots \cup A_n) = \sum_i P(A_i) - \sum_{i \neq j} P(A_i)P(A_j) + \sum_{i \neq j \neq k} P(A_i)P(A_j)P(A_k) \ldots$$

An important result that follows from the two-event case is as follows.

1. If A, B are mutually exclusive but one or other must occur

$$P(A) + P(B) = P(A \cup B) = 1$$

2. If B is the converse of A, i.e. B is 'A does not occur', written $B = \text{not} - A$, or $B = \neg A$, then

$$P(A) + P(B) = 1$$

and so

$$P(\neg A) = 1 - P(A)$$

6.1.1 Total probability, Bayes' theorem

Suppose a set of events E is made up of a number of non-overlapping sets E_i, $i = 1, 2, \ldots, n$; i.e.

$$E = \{E_i\}$$

where $E_i \cap E_j = \phi$ (the empty set) for all i, j with $i \neq j$, and that an event B depends on at least one of the E_i; then from what we have just shown

$$P(B) = P(B|E_1)P(E_1) + P(B|E_2) + \ldots + P(B|E_n)$$

$$= \sum_i P(B|E_i)P(E_i)$$

Now if E_j is any one of the E_i

$$P(E_j \cap B) = P(E_j|B)P(B)$$

Therefore

$$P(E_j|B) = \frac{P(E_j \cap B)}{P(B)} = P(E_j)\,\frac{P(B|E_j)}{P(B)}$$

This is Bayes' theorem:

$$P(E_j|B) = \frac{P(E_j)P(B|E_j)}{\sum_i P(E_i)P(B|E_i)}$$

The importance of this theorem is that if the events E_i are the possible causes of the effect B it enables us to calculate the probability that the cause was the particular event E_j when B was observed to occur.

6.2 PROBABILITY LAWS

A number of laws describing probabilities are needed in discussing and measuring quality. They fall into two classes according to whether they concern discrete or continuous phenomena: the first relates to events that can be counted, such as the number of machine breakdowns during a given period; the second relates to measurements of physical quantities such as length, weight or electrical resistance.

6.2.1 Discrete laws

(a) *Binomial law*

The binomial law concerns sampling from a batch (of manufactured items for example) when the composition of the batch is not altered by the drawing of the samples; this will be the case when the sample size is small compared with the batch or when the sample is returned to the batch after examination—this is called non-exhaustive sampling. If samples of n items are drawn from a batch of N the condition for applicability of the binomial law is that $n \leqslant N/10$.

If p is the fraction of defective items in a batch (the percentage is $100p$) the binomial law $B(n, p)$ states that the probability that a random sample of n will contain exactly k defectives is

$$P(x = k) = \binom{n}{k} P^k(1 - p)^{n-k}$$

where

$$\binom{n}{k} = \frac{n!}{k!(n - k)!}$$

The characteristics of $B(n, p)$ are as follows.

$$E(x) = np$$

where $E(x)$ is the mathematical expectation. This means that if the sampling is repeated many times the average number of defectives will be np. The variance is given by

$$\sigma^2(x) = E(x - np)^2 = np(1 - p)$$

It is conventional to write $q = 1 - p$ (q is the fraction of non-defective items) and so the variance is npq.

The distribution is illustrated graphically in Fig. 6.1.

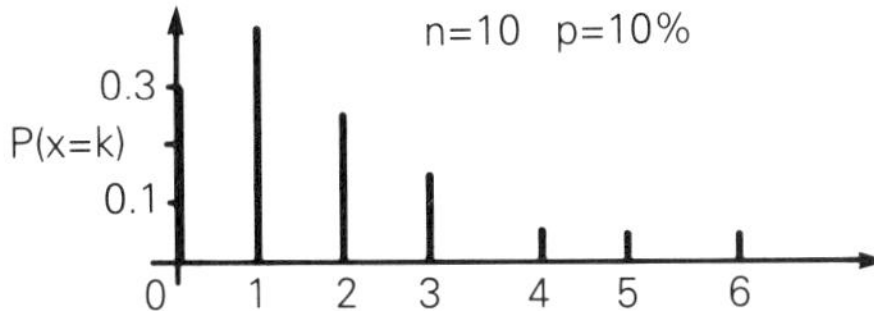

Figure 6.1 Binomial law: probability distribution.

The *cumulative function* $F(k)$ is the probability $P(x \leqslant k)$ that the sample will contain at most k defectives:

$$P(x \leqslant k) = \sum_{j=0}^{k} \binom{n}{j} p^j(1 - p)^{n-j}$$

This is illustrated in Fig. 6.2. Tables are available for $P(x = k)$ and $P(x \leqslant k)$.

Example 1

A piece of electronic equipment requires four resistors; these are drawn from a large batch for which it is known that the fraction of defectives is 5%. What is the probability that (1) three of the four will be defective and (2) at most three will be defective?

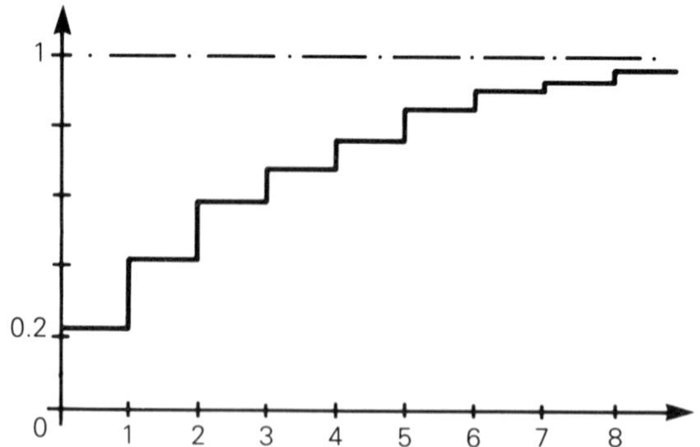

Figure 6.2 Binomial distribution: cumulative distribution.

(1) The probability of three defectives is

$$\binom{4}{3}(0.05)^3(1 - 0.05) = 0.0005$$

(2) The probability of at most three defectives can be calculated by summing the probabilities for 0, 1, 2, 3 defectives, since these are mutually exclusive events; but since there must be either 0, 1, 2, 3 or 4 in the sample the probability is $P(x \leqslant 3) = 1 - P(x = 4) = (1 - 0.05)^4 =$ 0.9999.

(b) *The hypergeometric law*
The hypergeometric law replaces the binomial law when the assumptions on which the latter are based cannot be made and the sampling does affect the composition of the batch. If, as before, the batch and sample sizes are N and n respectively, p is the fraction of defectives in the batch and $q = 1 - p$, the result is

$$P(x = k) = \binom{Np}{n}\binom{Nq}{n - k}\bigg/\binom{N}{n}$$

The characteristics are as follows:

$$E(x) = np$$

(as for the binomial) and

$$\sigma^2(x) = \frac{N - n}{N - 1}\,npq$$

The cumulative function

$$F(k) = P(x \leqslant k) = \sum_{j=0}^{k} P(x = j)$$

This distribution has the same graphical appearance as the binomial.

Example 2
A batch of 25 items is known to contain five defectives. What is the probability that a sample of five will contain three defectives?

The answer is

$$P(x = 3) = \binom{5}{3}\binom{20}{2}\Big/\binom{25}{5} = 0.0357$$

(c) *The Poisson law*
This important law can be regarded in either of two ways:

1. as a limiting form of the binomial when the batch size N is effectively infinite, the sample size n is very large and the probability p (here of defectives) is very small but the product np has a finite value, m say;
2. as a description of the occurrence events, e.g. if a machine might break down at any time in a certain interval and the average number of failures in an interval of that length is known, the law gives the probabilities of 0, 1, 2, . . . failures occurring in the interval.

The Poisson law is that if m is the average number of defectives in the sample of the size that is to be drawn (or the average number of failures in the interval of interest) then the probability that the sample will contain k defectives (or that there will be k failures in the interval) is

$$P(x = k) = \exp(-m)m^k/k!$$

The characteristics are

$$E(x) = m$$

and

$$\sigma^2(x) = m$$

(the equality of the mean and variance is a strong characteristic of the Poisson distribution).

Graphically this again is similar to the binomial.

Standard tables of the Poisson distribution $P(x = k)$ and of the cumulative distribution $F(k) = P(x \leqslant k)$ are available.

Example 3
It is planned to give a one-day demonstration of a certain machine for which the average number of breakdowns in a five-day week is known to be 10. What is the probability that it will not fail during the demonstration?

Here the mean rate is $m = 10/5 = 2$ breakdowns per day and so the probability or no failures in a day is

$$P(x = 0) = \exp(-2)\, 2^0/0! = \exp(-2) = 0.135$$

6.2.2 Continuous laws

Here we are dealing with variables measured on continuous scales and are concerned with such things as the probability that the value of a quantity x lies in a certain range or does not exceed a certain limit. Corresponding to the probability function $P(x = k)$ of the discrete laws we now have $p(x, x + \mathrm{d}x)$, the probability that the value of a random variable X lies between x and $x + \mathrm{d}x$, where usually $\mathrm{d}x$ is small. We write

$$p(x, x + \mathrm{d}x) = f(x)\,\mathrm{d}x$$

where $f(x)$ is the *probability density function*. The corresponding cumulative function $F(k)$, the probability that the value of x does not exceed k, is

$$F(k) = P(x \leqslant k) = \int^{k} f(x)\,\mathrm{d}x$$

where the lower limit for the integral depends on the range of values that x can take; this can be 0 or $-\infty$ or some finite non-zero value.

(a) *The normal (Gaussian) law*
For a random variable with mean m and standard deviation σ this is

$$f(x) = \frac{1}{\sigma\sqrt{2\pi}} \exp\left[-\frac{(x - m)^2}{2\sigma^2}\right]$$

with characteristics

$$E(x) = m$$

and

$$\sigma^2(x) = \sigma^2$$

Figure 6.3 gives a graph of $f(x)$; it is symmetrical about $x = m$, i.e. $f(m - x) = f(m + x)$.

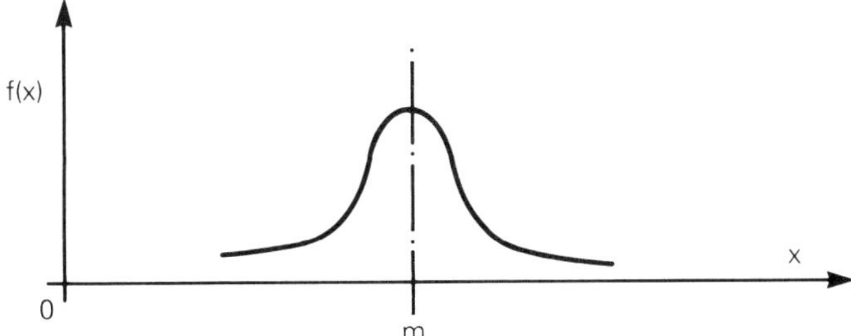

Figure 6.3 Gaussian (normal) law: probability distribution.

Quantities that are important in the use of this distribution are the fractions of the total population in intervals $m \pm k\sigma$, i.e. within k standard deviations on either side of the mean, for various values of k. The values are

$$k = 1 \quad 68.26\%$$

$$k = 2 \quad 95.45\%$$

$$k = 3 \quad 99.73\%$$

These are shown in Fig. 6.4. The cumulative function is

$$F(x) = \int_{-\infty}^{x} f(\xi)\, \mathrm{d}\xi$$

where the lower limit is $-\infty$ because there is no restriction on the range of values that x can take.

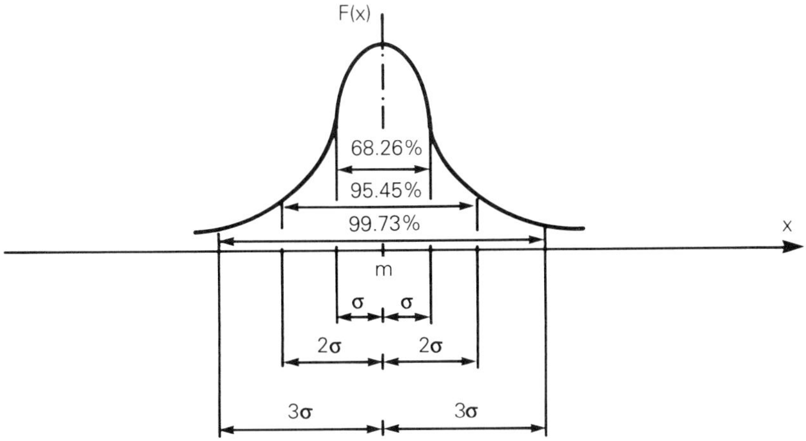

Figure 6.4 Gaussian (normal) law: critical ranges for distance from the mean.

It follows from the form of $f(x)$ that $F(+\infty) = 1$ (as it should be) and that $F[(x - m)/\sigma] = 1 - F[(m - x)/\sigma]$. $F(x)$ is shown in Fig. 6.5.

The integral for $F(x)$ cannot be evaluated analytically, and so numerical methods such as Simpson's rule have to be used. In practice standard computer programs are now used, or the standard tables that are available. The tables give the value as a function of the 'reduced' variable $u = (x - m)/\sigma$, for which, for example, $F(-u) = 1 - F(u)$. Such a table is given in Appendix 1.

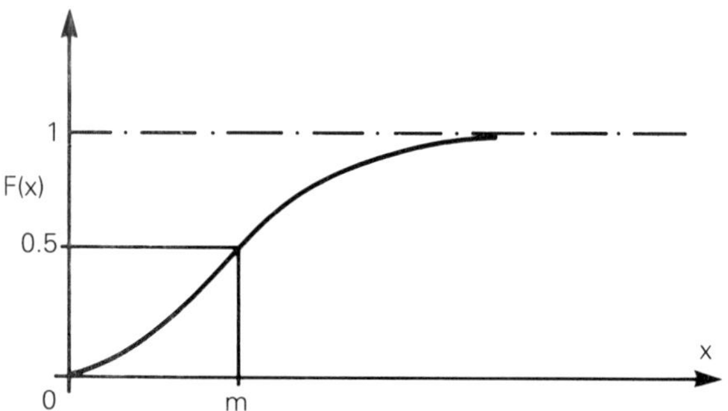

Figure 6.5 Gaussian (normal) law: cumulative distribution.

Example 4

The resistances of a production batch of resistors are normally distributed with mean 600 mΩ and standard deviation 120 nΩ. For a certain assembly the permissible upper and lower limits are 720 and 420 mΩ respectively. What percentage of the batch will meet this requirement?

The fraction is

$$\int_{420}^{720} \frac{1}{120\sqrt{(2\pi)}} \exp\left[-\frac{(x - 600)^2}{2\sigma^2}\right] dx = F(u_2) - F(u_1)$$

where

$$u_2 = (720 - 600)/120 = 1$$

$$u_1 = (420 - 600)/120 = -1.5$$

From the table we find $F(1) = 0.8413$ and $F(-1.5) = 1 - F(1.5) = 0.0668$ and so the fraction is $0.8413 - 0.0668 = 0.7745 = 77\%$.

6.2.3 Central limit theorem, distribution of the mean

This important theorem states that:

> if $X_1, X_2, \ldots, X_n$ are independent random variables with arbitrary distribution laws then the distribution of the sum $Y = \Sigma_i X_i$ tends to the normal law as n increases.

The justification of the use of the normal distribution in many applications of statistical tests rests on this theorem. It is important that the influence of each X_i is small and that all have more or less equal influences.

Whatever the distributions of the X_i

$$\bar{Y} = \sum_{i=1}^{n} \bar{X}_i$$

and $\sigma^2(Y) = \Sigma\, \sigma^2(X_i)$.

We are interested especially in the distribution of the mean of some property measured for each item in a sample of n. If these values are distributed normally with different means but all with the same standard deviation, the mean also is distributed normally, with parameters

$$\bar{X} = \frac{1}{n} \sum_{j=1}^{m} \bar{X}_j \qquad \sigma(\bar{X}) = \frac{\sigma(x)}{\sqrt{n}}$$

A corollary is that the difference Z of two normally distributed random variables X, Y is distributed normally with parameters

$$\bar{Z} = \bar{X} - \bar{Y} \qquad \sigma(Z) = [\sigma^2(X) + \sigma^2(Y)]^{1/2}$$

6.2.4 The log normal law

Here the logarithm of the random variable x is distributed normally; if as usual the parameters are m and σ the probability distribution function is

$$f(x) = \frac{1}{\sigma\sqrt{(2\pi)}}\frac{1}{x}\exp\left[-\frac{(\ln x - m)^2}{2\sigma^2}\right] \quad \text{for } x \geqslant 0$$

$$= 0 \qquad\qquad x < 0$$

The characteristics are:

$$E(x) = \exp(m + \tfrac{1}{2}\sigma^2)$$

$$\sigma^2(x) = \exp(2m + \sigma^2)[\exp(\sigma^2) - 1]$$

The cumulative function $F(x)$ is computed by changing to the reduced variable $u = (\ln x - m)/\sigma$. Figure 6.6 illustrates $f(x)$ and $F(x)$.

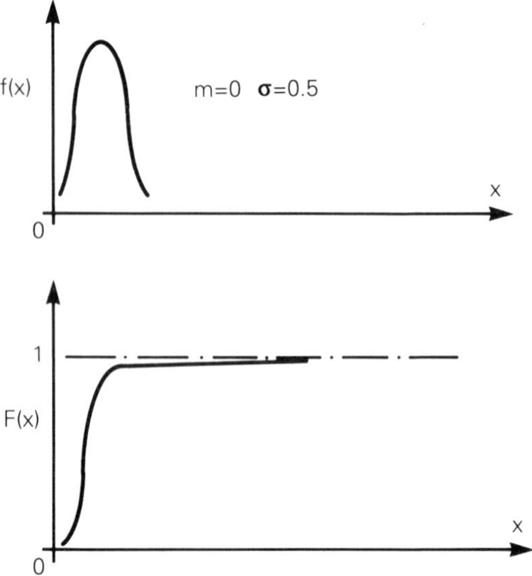

Figure 6.6 Log normal law: (a) probability (b) cumulative distribution.

6.2.5 The exponential law

This is particularly relevant to the reliability of electronic equipment; like the Poisson law for discrete variables it describes events that can be considered as occurring at random, such as breakdowns.

The probability density function is

$$f(x) = \lambda \exp(-\lambda x)$$

for $x \geqslant 0$. The characteristics are

$$E(x) = 1/\lambda$$

$$\sigma^2(\chi) = 1/\lambda^2$$

The cumulative function is

$$F(x) = \int_0^x f(\xi)\,\mathrm{d}\xi = 1 - \exp(-\lambda x)$$

Figure 6.7 illustrates $f(x)$ and $F(x)$.

This law is closely related to the Poisson law. Let λ be the average rate at which certain events such as breakdowns occur, so that the average number in a period of length x is λx; then if in the Poisson law the parameter m has the value λx the probability of k breakdowns in the period x is

$$P(k) = \exp(-\lambda x)(\lambda x)^k/k!$$

and the probability of no breakdowns in that period is

$$P(0) = \exp(-\lambda x)$$

which is the exponential law.

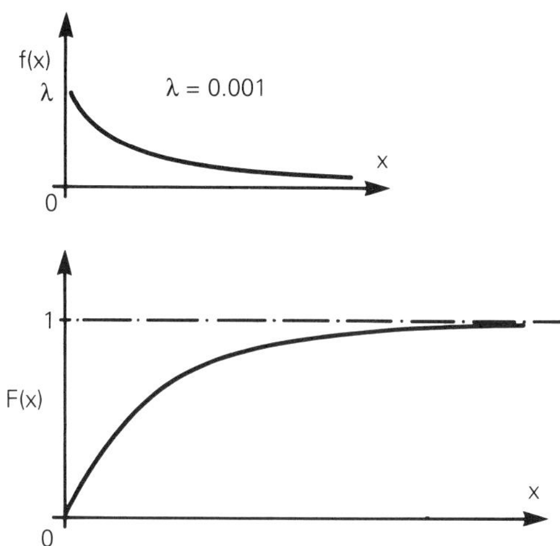

Figure 6.7 Exponential law: (a) probability (b) cumulative distribution.

Example 5
The average failure rate of a piece of electronic equipment is estimated as 1 failure per 100 000 h (10^{-5} failures per hour). What is the probability that it will fail between 200 and 300 h in operation?

Here the variable x of the law is the time t; the probability that the equipment will fail between 200 and 300 h is

$$F(300) - F(200) = \exp(-0.002) - \exp(-0.003) = 0.001$$

6.2.6 The Weibull law

This is much used in reliability studies, particularly for mechanical systems. It has the advantage of being very flexible and can be adapted to the needs of a variety of circumstances.

The probability density function involves three parameters:

$$f(x) = \frac{\beta}{\eta}\left(\frac{x-\gamma}{\eta}\right)^{\beta-1}\exp\left[-\left(\frac{x-\gamma}{\eta}\right)^{\beta}\right]$$

where $x - \gamma > 0$, β is the shape parameter (a dimensionless number), η is the scale parameter (dimension of the variable x (here, time)) and γ is the location parameter, also of dimension x. The cumulative function is

$$F(x) = 1 - \exp\left[-\left(\frac{x-\gamma}{\eta}\right)^{\beta}\right]$$

The characteristics are

$$E(x) = \gamma + \eta\Gamma(1 + 1/\beta)$$

where Γ denotes the gamma function (see Appendix 9) and

$$\sigma^2(x) = \eta^2\left\{\Gamma\left(1 + \frac{2}{\beta}\right) - \left[\Gamma\left(1 + \frac{1}{\beta}\right)\right]^2\right\}$$

Figure 6.8 illustrates $f(x)$ and $F(X)$.

6.3 CONFIDENCE INTERVAL FOR THE MEAN

6.3.1 When the variance is known

If $\bar{X}$ is the mean of a set of n normal random variables X_i, $i = 1, 2, \ldots, n$, all of which have mean m and standard deviation $\sigma(X)$, i.e.

$$\bar{X} = \frac{\Sigma_i X_i}{n}$$

then $\bar{X}$ is distributed normally with mean m and standard deviation $\sigma(\bar{X}) = \sigma(X)/\surd n$. The *symmetric confidence interval* at probability level α is defined as

$$\bar{X} - u(1 - \tfrac{1}{2}\alpha)\frac{\sigma}{\surd n} < m < \bar{X} + u(1 - \tfrac{1}{2}\alpha)\frac{\sigma}{\surd n}$$

where $u(1 - \frac{1}{2}\alpha)$ is the value of the reduced normal variable for the argument $1 - \frac{1}{2}\alpha$.

6.3.2 When the variance is not known

This has now to be estimated from the sample; the unbiased estimate is

$$\hat{\sigma}^2(\bar{X}) = \frac{\Sigma_i(X_i - \bar{X}^2)}{n-1}$$

It can be shown that $(\bar{X} - m)/(\sigma/\surd n)$ is distributed as Student's t with $n - 1$ degrees of freedom. It follows that the corresponding confidence interval is now

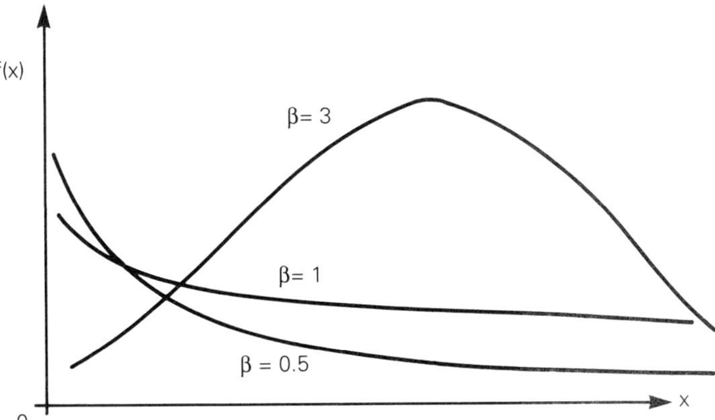

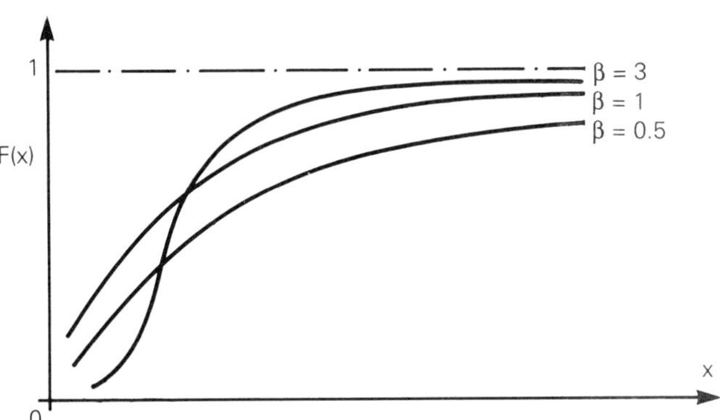

Figure 6.8 Weibull law: examples of (a) probability and (b) cumulative distribution.

$$\bar{X} - t(1 - \tfrac{1}{2}\alpha; n - 1)\frac{\sigma}{\sqrt{n}} < m < \bar{X} + t(1 - \tfrac{1}{2}\alpha; n - 1)\frac{\sigma}{\sqrt{n}}$$

6.3.3 Confidence interval for the mean time between failures

This derivation is based on the assumption of truncated sampling with replacement of failed items. If n is the number of items tested, t is the duration of the test and n is the number of failures recorded, then the estimate of MTBF is

$$\widehat{\text{MTBF}} = nt/r$$

An interval AB, where A and B are the lower and upper bounds respectively, can be determined such that the probability that the MTBF lies in AB is $1 - \alpha - \beta$. A and B depend on α and β and are given by

$$A = \frac{2nt}{X^2(1 - \beta; 2r + 2)} \qquad B = \frac{2nt}{X^2(\alpha; 2r)}$$

For any particular case these bounds can be either calculated from tables of the χ^2 distribution (e.g. Appendix 3) or read from the graphs of Fig. 6.9.

Example 6

$$n = 1000,\ r = 4,\ t = 100\ \text{h};\ \alpha = \beta = 0.05.$$

The estimated MTBF is 1000 × 100/4 = 25 000 h.

$$\text{Lower limit} = 0.39 \times 25\,000 = 9750\ \text{h}$$

$$\text{Upper limit} = 3.8 \times 25\,000 = 95\,000\ \text{h}$$

6.4 LINEAR REGRESSION

The problem here is that we have a set of pairs of observed values (x_i, y_i) – a cluster of points when plotted – and that we would like to represent them as well as possible by a straight line, i.e. to fit them to a model (Fig. 6.10)

$$y = ax + b$$

We have to find values for a and b. The procedure used in regression analysis is the method of least squares. This finds the values of a and b that minimize the sum of the squares of the differences between the observed values y_i and the values of $ax_i + b$ given by the model. These differences $y_i - ax_i - b$ are the 'errors' e_i in the model's predictions. The method minimizes $E = \sum_i (e_i)^2$.

We have

$$E = \sum (y_i - ax_i - b)^2$$

$$= \sum (y_i)^2 - 2a \sum x_i y_i - 2nb\bar{y} + a^2 \sum (x_i)^2 + 2nab\bar{x} + nb^2$$

where n is the number of pairs of (x_i, y_i). The values of a and b that minimize E are given by the solutions of

$$\frac{\partial E}{\partial a} = 0 \qquad \frac{\partial E}{\partial b} = 0$$

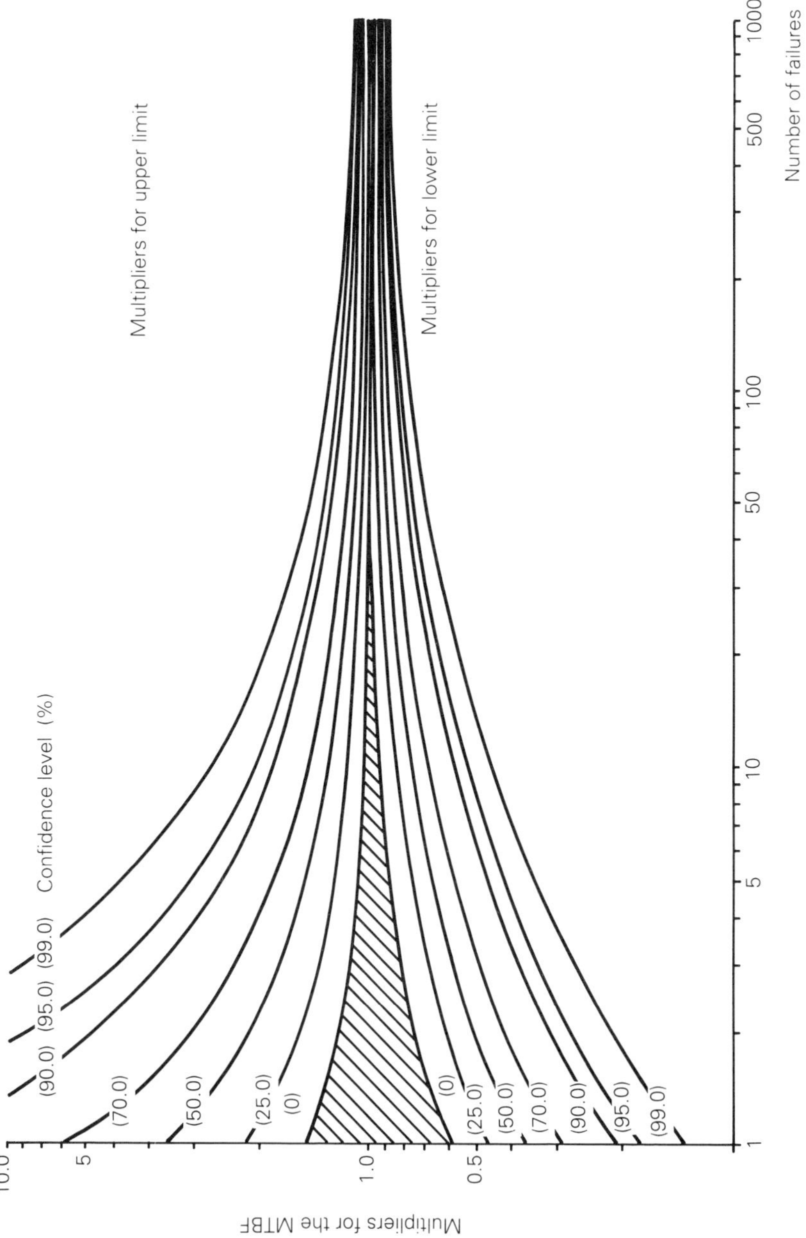

Figure 6.9 (a) Confidence intervals, truncated sampling: factors by which the estimated MTBF should be multiplied to give upper and lower limits at a stated level of confidence.

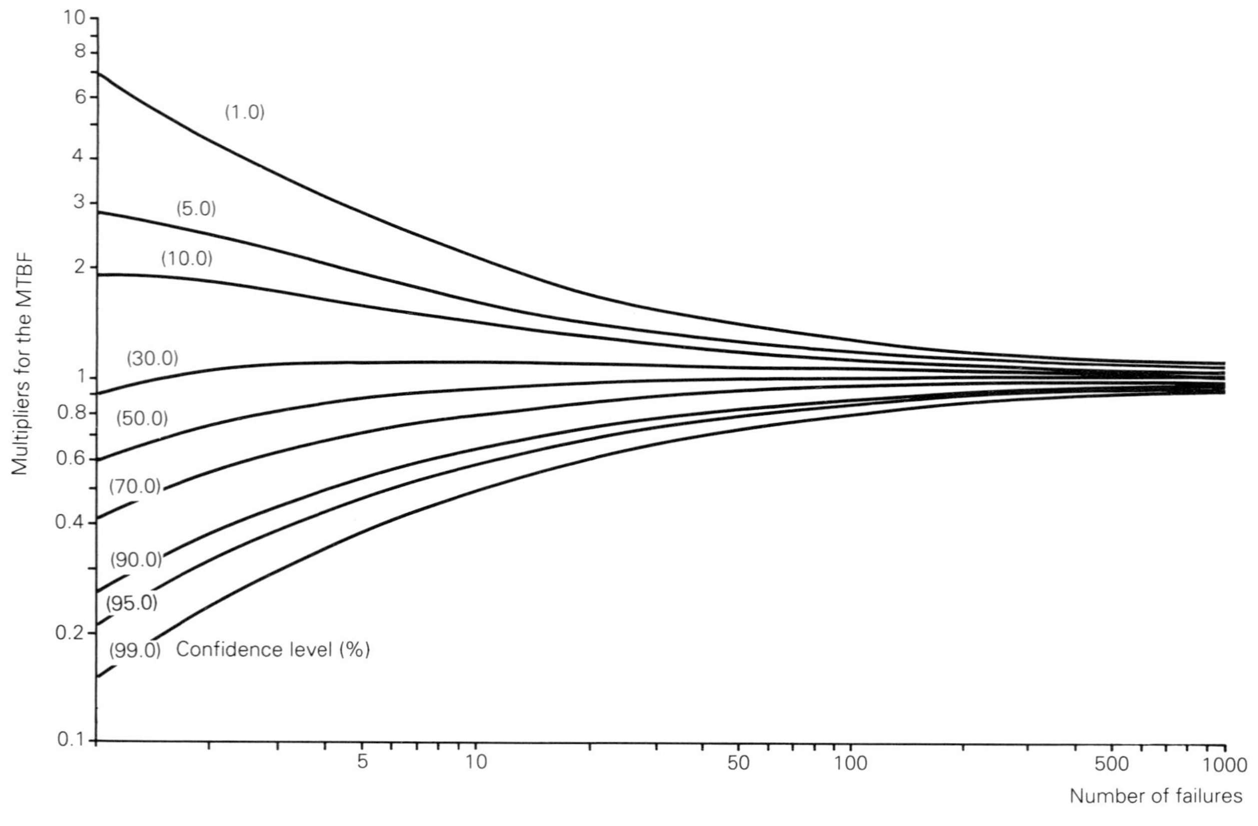
Multipliers for the MTBF
10
8
6
4
3
2
1
0.8
0.6
0.4
0.2
0.1
(1.0)
(5.0)
(10.0)
(30.0)
(50.0)
(70.0)
(90.0)
(95.0)
(99.0)
Confidence level (%)
5
10
50
100
500
1000
Number of failures

Figure 6.9b

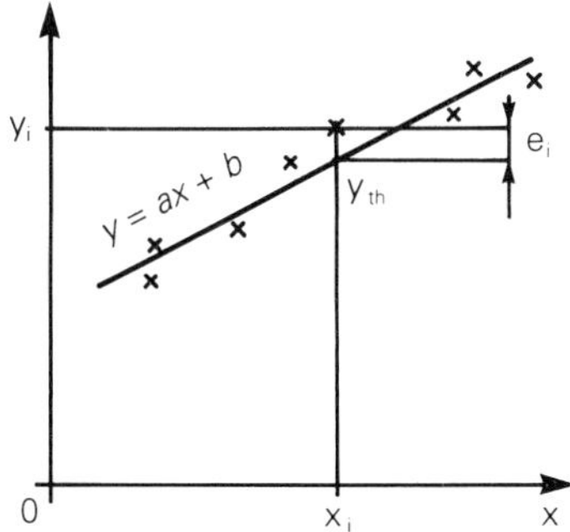

Figure 6.10 Linear regression.

Differentiating and solving the resulting linear equations gives

$$a = \frac{\Sigma x_i y_i - n\bar{x}\bar{y}}{\Sigma (x_i)^2 - n\bar{x}^2} \qquad b = \bar{y} - a\bar{x}$$

This gives the 'best' line in the least-squares sense.

The method can be used for other models. Thus for the quadratic model $y = ax^2 + bx + c$ we find E, the sum of the squares of the errors e_i, as before and solve $\partial E/\partial a = 0$, $\partial E/\partial b = 0$ and $\partial E/\partial c = 0$ for a, b and c.

Standard programs are available for fitting straight lines and other curves to statistical data.

Example 7
For the values

x	20	70	110	160	190
y	60	90	120	140	170

the method gives $y = 0.62x + 48$.

Exercises

(Solutions are given on pp. 148–153)

1. An automatic machine makes spacer bars whose length must be between 37.45 and 37.55 mm; the lengths produced have a normal distribution with mean 37.50 mm.

   (i) What must be the standard deviation if 998 out of every 1000 bars are to be acceptable?
   (ii) A random sample is drawn from the production and the lengths are measured. What must be the size of this sample if the mean of the lengths is to lie between 37.495 and 37.505 with probability 0.95?

   You are given that if z is the reduced central normal variable

$$P(0 \leqslant z \leqslant 1.96) = 0.475$$

$$P(0 \leqslant z \leqslant 2.05) = 0.480$$

$$P(0 \leqslant z \leqslant 3.10) = 0.499$$

2. An automatic machine makes items whose weight is distributed normally with mean 0.90 g and standard deviation 0.06 g.

   (i) What is the probability that the weight of an item chosen at random lies between 0.84 and 0.99 g?
   (ii) How many items of weight less than 0.81 g can one expect there to be in a batch of 5000?
   (iii) The items are packed into boxes, 100 to a box, by another machine; a certain number of boxes are chosen at random and the mean weight w of an item is found for each box.

      (a) What are the mean and standard deviation of w?
      (b) What is the probability of a measurement deviating by 1% from this mean value?
      (c) What are the 95% confidence limits for the mean item weight in a box of 100?

(d) The mean weight of the items in a box of 100 chosen at random is found to be 0.88 g. Can the box be considered as representative at the 1% level?

3. The lifetime, in hours, of an electric light bulb is a normal variable of mean M and standard deviation 20. A test of a sample of 16 gives a mean life of 3000 h. Find the 90% confidence interval for M.

4. We wish to calculate the probability that a structural element will break under deflection. The method employed is what is called the R/C – resistance/constraint – method and involves constructing the random variable $R - C$, from which the probability can be deduced.

   The resistance R is normally distributed with mean $\bar{R} = 28$ and standard deviation $\sigma_R = 2$.

   The constraint C is normally distributed with mean $\bar{C} = 25$ and standard deviation $\sigma_C = 1.5$.

   (i) Give the theoretical calculation.
   (ii) Find the probability of breaking. Units are daN/mm^2 (decanewtons per square millimetre).

5. A machine makes items whose diameter X is a normal random variable with mean 32 mm and standard deviation 1 mm.

   (i) What is the probability that the diameter of an item is less than 30.5 mm?
   (ii) What is the probability of a diameter between 31 and 33 mm?
   (iii) As a control, samples of 20 are taken at regular intervals and measured; if the mean diameter found is $\bar{X}$

   (a) what is the probability distribution of $\bar{X}$?
   (b) in what interval $[a, b]$ must $\bar{X}$ lie if the machine can be considered to be correctly adjusted with a probability of 0.99?

6. A factory makes a certain item in large numbers. This is done in two stages, in the first of which a defect A can appear and in the second a defect B. Experience has shown that 2% of the items show defect A and 8% show B.

   (i) Find the probability that an item chosen at random

   (a) has both defects,
   (b) has at least one of the defects,
   (c) has one and only one defect,
   (d) has no defect.

(ii) A sample of 200 items is taken and the number X showing fault A is noted.

(a) X is regarded as a random variable with Poisson distribution. What justification is there for this? What is the parameter?
(b) What is the probability that 10 items in the sample of 200 will show fault A?

(iii) A sample of 300 is taken and the number Y showing fault B is noted.

(a) Y is regarded as a normally distributed random variable. What are the parameters?
(b) Calculate $\text{prob}(Y < 24)$
$\text{prob}(20 < Y < 35)$
$\text{prob}(Y < 30, \text{ given that } Y > 24)$

7. The objective of this problem is the investigation of the performance of a lathe, one of the machines in a company's mechanical workshop. The problem is in two independent parts.

The lathe turns shafts to a nominal diameter of 24 mm; the actual diameter is a normal random variable of mean 24 and standard deviation 0.02 mm, and the tolerance limits are 23.95, 24.05.

(i) How many good items will there be in a sample of 1000?
(ii) As a control, 20 items are chosen at random and the mean diameter $\bar{X}$ is found; what is the interval $[a, b]$ in which this must lie if the lathe can be regarded as correctly adjusted with probability 0.99?
(iii) A check on 20 items gives the following values:

Diameter	Number	Diameter	Number
23.93–23.95	1	24.01–24.03	8
23.95–23.97	1	24.03–24.05	2
23.97–23.99	1	24.05–24.07	0
23.99–24.01	7		

(a) Find the mean and standard deviation of this set of values.
(b) What conclusion do you draw from these measurements?

8. A quality assurance service decides to impose controls on dimensions X and Y of a product, using control charts for the mean and standard deviation. It has to estimate the population standard deviation σ, and has the following observations:

X			
15.050	15.049	15.052	15.056
15.055	15.058	15.061	15.058
15.056	15.061	15.070	15.049
15.060	15.046	15.061	15.059
15.062	15.054	15.07	15.080

Y			
30.021	30.035	30.020	30.021
30.025	30.032	30.019	30.025
30.031	30.035	30.022	30.024
30.026	30.032	30.018	30.023
30.027	30.035	30.017	30.040

(i) Give the control limits for the two machines.
(ii) Set up the required control charts, for a sample size 5.

A study of the way the dimension X changes with time enables the effect of wear to be estimated, and consequently the time the machine tool can be allowed to run before there is a risk of rejects being produced. Observation gave the following:

Operating time	X
1 min	15.000
8 min	15.010
35 min	15.040
50 min	15.070
70 min	15.085
80 min	15.090

By fitting a straight line $X = at + b$ to these values find the intervals at which the machines should be re-set.

9. A manufacturer of food products is looking for the most economical way of putting a powder into cartons. Trials using an old machine gave the following results:

Sample no.			Weight (g)		
1	510	522	520	514	516
2	514	516	512	514	520
3	516	514	518	512	514
4	510	508	520	516	514
5	516	504	512	516	522
6	518	510	512	518	514
7	512	512	508	512	520
8	514	518	514	512	516
9	518	517	515	514	510
10	520	515	514	508	513
11	518	514	516	512	516
12	510	511	512	512	510
13	520	518	506	518	510
14	518	515	516	512	512
15	522	506	510	522	522
16	516	516	514	510	516
17	516	510	516	520	522
18	514	512	514	518	512
19	524	502	516	520	508
20	514	514	524	516	518

Here each observation is the total weight of the box and its filling; it is known that the box weight is distributed normally with mean 64 g and standard deviation 0.05 g.

(i) (a) Can such measurements be used to keep the production under statistical control? Use either a goodness-of-fit test or Henry's line. What are the relative advantages and disadvantages of the two methods?

(b) Give the different estimators that can be used for σ.

(ii) The customer requires that the average content of a box shall be not less than 445 g and that at most 2% can be more than 7 g short.

(a) What percentage of boxes will contain less than the minimum weight stipulated by the customer?

(b) If the mean is changed what must its new value be to stay within the terms of the contract?

(iii) Set up the control chart that must be used for ensuring that (ii) is satisfied, assuming a sample size of 5.

Samples of the output were taken at different times and the following weights were found:

at	10.00 h	516,	514,	512,	520,	514
	10.30	514,	518,	510,	520,	512
	11.00	510,	516,	520,	514,	522
	11.30	522,	504,	512,	502,	508

Incorporate these in your control chart.

(iv) What would be the cost to the manufacturer of adhering strictly to the customer's conditions ((ii)(a) above)? The annual sales are 5 hundred thousand boxes of 445 g, priced at £1.50 per kilogram gross weight.

(v) In response to invitations to tender for new packaging machinery the manufacturer has these proposals:

Machine A: $\sigma_A = 1$ g guaranteed
cost £100 000
estimated life 10 yr
residual value after 10 yr £9 500

Machine B: $\sigma_B = 3$ g guaranteed
cost £55 000
estimated life 10 yr
residual value after 10 yr £5 000

The current value of the existing machine is zero.

What should be the decision, on economic grounds?

10. A workshop has 30 workstations, each equipped with a machine A and a machine B. The two machines operate completely independently of each other. The probabilities of breakdown in the course of a working day are $P_A = 0.1$, $P_B = 0.03$.

(i) For an arbitrarily chosen workstation during a working day find:

(a) the probability P_1 that both A and B break down;
(b) the probability P_2 that at least one machine breaks down;
(c) the probability P_3 that A, and A alone, breaks down;
(d) the probability P_4 that one, and only one, machine breaks down.

(ii) For an arbitrarily chosen group of 10 machines of type A and a single working day let X be the number that break down during that day. Give the probability distribution of X.

(a) Find the probability that exactly three of these 10 machines fail.

(b) Find the probability that at least one fails.

Give these probabilities to an accuracy of 10^{-4}.

(iii) Let Y be the number of workstations at which one and only one machine breaks down during the working day. Justifying the use of the Poisson law, calculate:

(a) the probability that four of the 30 stations will experience a breakdown of one and only one machine;

(b) the probability that at most two stations will experience a breakdown of one and only one machine.

Give these probabilities to an accuracy of 10^{-2}; for (b) take $P_4 = 0.12$.

(iv) Two dimensions x, y are to be controlled for an item produced by the workstations, and the limits are $x = 6 \pm 0.01$, $y = 3 \pm 0.01$. Measurements on the production of a station give the mean and standard deviations: $E(x) = 6.003$, $\sigma_x = 0.005$; $E(y) = 3.002$, $\sigma_y = 0.005$.

Assuming the x, y are normally and independently distributed, what percentage of rejects can be expected in a day's output?

(v) Because of the high cost of maintenance of machine A the management decides to reconsider the maintenance policy. From records of the numbers of fault-free working days over a year it appears that the lifetimes of these machines follow a Weibull law with parameters $\gamma = 80$, $\eta = 110$, $\beta = 2.2$.

Let the random variable t be the number of fault-free working days before the first failure. On Weibull paper show the cumulative distribution function F for this variable, for $t \in [100, 250]$.

(a) Find the mean number of fault-free working days.

(b) Hence find the probability of not failing during a period of this length.

(c) Using the graph of F find the number of days after which 30% of the machines will have had their first failure.

11. A search through the maintenance file has revealed 27 failures of pumps of a certain make during the past 2 months. The times of fault-free operation, in increasing order, are as follows:

26	132	245	h
43.4	145	247	
58.8	151	282	
68	171	295	
80.5	180	307	
94	192	320	
98	202	350	
112	210	335	
125	220	474	

(i) Using Weibull paper and the table of the Weibull law provided (Appendix 5) find the MTBF: use the method of mean ranks to determine the cumulative relative frequenciees $F_i = (\Sigma_i n_i)/(N + 1)$.

(ii) Hence determine the maintenance routine best suited to this product. Justify your choice.

12. During the period 1.9.89 to 1.9.90 a certain machine has been in operation for 1205 days and has failed 10 times. The intervals between successive failures were as follows (in days):

44 68 82 39 108 299 57 255 151 49

(i) Tabulate these values in a manner that shows the numerical order of the failure, the times between successive failures in increasing order and the accumulated good time.

(ii) Let $R(t)$ denote the probability that the machine will run for a time t without failing. Using log–linear paper, justify the representation of $R(t)$ by the exponential law.

(iii) Find the MTBF for this machine.

13. In order to obtain information concerning the reliability of a certain machine the performance of the example installed by a customer has been studied, and the following intervals between successive failures have been recorded:

170 225 260 300 320 330 390 490 h

(i) Find the Weibull model.

(ii) Find the maintenance interval corresponding to a 50% risk of failure.

(iii) (a) Draw the curve of

$$\lambda(t) = \frac{\beta}{\eta}\left(\frac{t-\gamma}{\eta}\right)^{\beta-1}$$

for $t = [200, 400]$.

(b) Interpret your results.

14. (i) The following lifetimes were recorded for a certain device:

1860 2500 2900 3600 3950 5100 6300 h

(a) Are these described by a Weibull law?
(b) If so, what are the parameters?
(c) Compute $E(t)$.

(ii) A control scheme is defined as follows:

$$n = 20,\ A = 2,\ R = 3,\ \text{AQL} = 2\%,\ P_2 = 10\%$$

α, β denote the risks borne by the supplier and the customer respectively.

(a) Compute P_1, α, β.
(b) Draw the efficiency curve for the scheme.

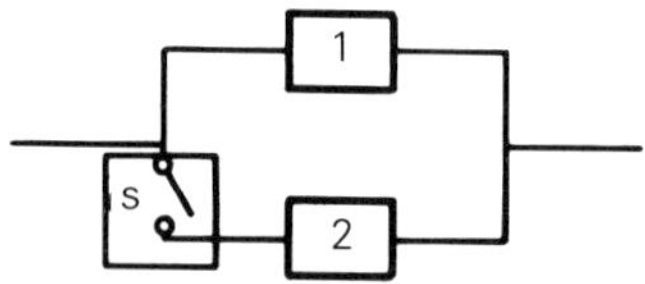

(iii) In the system of the diagram element 1 is working; if it fails it is switched out by S and element 2 is switched in. The following assumptions are made:

- switch S has reliability = 1;
- the repair time follows an exponential law with parameter N;
- the time of fault-free operation of each element follows an exponential law with parameter λ;
- at $t = 0$, $P(0) = 1$, $P(1) = P(2) = 0$.

We are interested in the availability of this system.

(a) Draw the Markov chain.
(b) Give the state-change equations.
(c) Apply the Laplace transform and give the resulting equations in matrix form.
(d) Compute the availability.
(e) What is the limiting value of this as $t \to \infty$?

(iv) The question concerns an aircraft blind-landing system.

The risk of failure for any single flight must be less than 10^{-6}. An aircraft has 10 essential (electronic) systems, for each of which the risk of failure in any one flight must be less than 10^{-7} (this is the risk of the complete blind-landing system failing). The landing system is in operation only during the 6-min approach to landing. A total of 1000 components are involved, for which the average value of λ is 10^{-5}/h.

(a) Find the probability that the system will fail during a flight.
(b) Does the system attain the specified risk level?
(c) If not, what should be done to achieve this?

15. To investigate the reliability of certain machines, a number of which are installed in a workshop and operated under the same conditions, the number of days of operation before the first failure have been recorded for 10 of the machines:

80 110 68 86 100 61 120 94 135 74

(i) Using Weibull paper, justify the representation of these times by a Weibull law with $\gamma = 50$. Find the other parameters β, η and the MTBF.
(ii) Find the running time for which the risk of failure is 50%.
(iii) Given that the failure rate is $\lambda(t) = (\beta/\eta)[(t-\gamma)/\eta]^{\beta-1}$, evaluate $\lambda(t)$ for $t = 60$, 80, 100, 120, 130 and plot these values. Interpret the result.

16. In a study to find estimates $\hat{F}(t)$, $\hat{R}(t)$ and $\hat{\lambda}(t)$ for the reliability functions for a certain device the times to failure of 55 examples were recorded, with the following results:

Time interval $t_i - t_{i+1}$	No. failing in this interval
0–500	3
500–1000	8
1000–1500	10
1500–2000	12
2000–2500	7
2500–3000	8
3000–3500	7

(i) Give the estimates.
(ii) Draw the graph of $\lambda(t)$. What do you conclude from this?

17. In a food products factory the conditioning operations and the handling of the products preparatory to despatch are automated.

This automation requires the inclusion of 80 extra devices, all of the same type and all subject to very heavy use – 1.5×10^6 handling cycles each per month. The quality engineer, in order to set criteria for reliability, has recorded the failures of a sample of 50 of these devices, with the following results:

Handling cycles	No. of failure
0 – 1.5	0
1.5– 3	1
3 – 4.5	2
4.5– 6	3
6 – 7.5	4
7.5– 9	5
9 –10.5	5
10.5–12.0	6
12.0–13.5	5
13.5–15	5
15 –16.5	4
16.5–18	3
18 –19.5	2
19.5–21	2
21 –23.5	1
23.5–24	1
24 –25.5	0
25.5–27	1

In general, the Weibull law is applicable to this type of equipment.

(i) Show how the special Weibull paper can be used to test the validity of the Weibull law in this case. Expressing the law in the standard form

$$R(t) = \exp\left[-\left(\frac{t-\gamma}{\eta}\right)^{\beta-1}\right]$$

find the values of β, γ, η.

(ii) Having found these values you wish to be assured that the Weibull law really does describe the data, and for this you apply the χ^2 test. How do you go about this? Give the full details of the calculation, and apply the test at the level $\alpha = 0.05$. [Table of χ^2 provided in Appendix 3]

18. A company that manufactures shipboard electric motors for the Navy must, without fail, ensure that the noise level of its products does not exceed a certain value (given in decibels), so that it is not detectable by enemy submarines. Investigation by the research department has shown how to deal with the well-known causes of excessive vibration (e.g. unbalanced rotor), but there are other factors whose effects are less easily identified:

Factor A: viscosity of lubricant, $A1$ = low, $A2$ = high
Factor B: method of tightening the cover,
$B1$ = uncontrolled, $B2$ = with a torque wrench
Factor C: play in bearings, $C1 < 6\ \mu m$, $C2 > 6\ \mu m$

To investigate these a $2 \times 2 \times 2$ experiment was undertaken. The Yates table for the results was as follows:

(1)	170
a	145
b	73
ab	167
c	179
ac	145
bc	78
abc	167

Use the Student test with confidence level 90% to find which factors have a significant effect.

19. (i) Find the overall reliability of the following system, giving $R(i) = 0.9$. Compare the result with that for a system without redundancy.

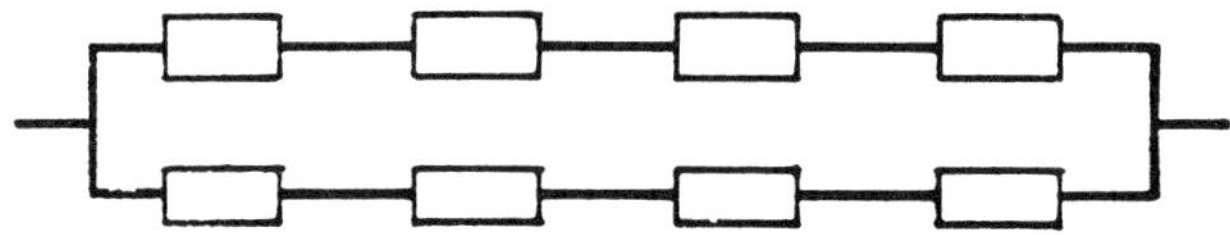

(ii) Repeat the calculation for the following system, with $R(i) = 0.9$ also.

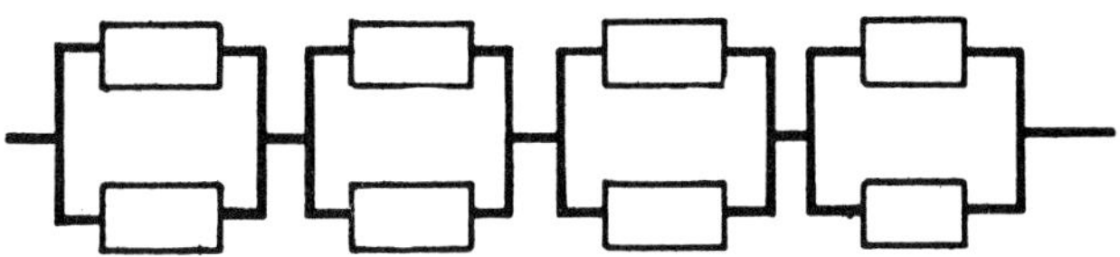

Solutions

1

(i) $\sigma = 0.016$
(ii) $n = 40$

2

(i) 77.4%
(ii) 334
(iii) (a) $\bar{X}_p = 0.90$ g; $\sigma_p = 0.006$ g
(b) $P = 0.88$
(c) No

3

$-1.64 \leqslant M \leqslant 1.64$

4

(i) P rupture $= P(R < C) = P(R - C < 0)$
$R - C$ is normally distributed with mean $m = \bar{R} - \bar{C}$, standard deviation

$$\sigma = (\sigma_C^2 + \sigma_R^2)^{1/2}$$

whence

$$(P(R - C < 0) = P\left(U < \frac{\bar{R} - \bar{C}}{\sigma}\right)$$

(ii) $P(U < 3/2.5) = P(U < -1.2)$

$$P(R - C < 0) = 0.12$$

5

(i) 6.68%
(ii) 68.26%
(iii) (a) $\bar{X}$ is normally distributed:

$N\left(\bar{X}\ \frac{\sigma_X}{\sqrt{n}}\right)$

(b) The interval is AB

$31.42 < \bar{X} < 32.57$

6

(i) (a) $P = 0.16\%$

(b) $P = 9.84\%$

(c) $P = 9.68\%$

(d) $P = 9.52\%$

(ii) $m = 4$

$P(A) = 0.53\%$

(iii) n is large

$m = np = 24$

$\sigma = 4.7$

$P(Y < 24) = 0.41\%$

$P(20 < Y < 35) = 26.4\%$

$P(25 < Y < 29) = 27.22\%$

7

(i) 986

(ii) [23.948 − 240.56]

(iii) (a) $\bar{X} = 24.006$; $\sigma = 0.028$

(b) Adjustment is correct

8

(i) For X:

$\sigma = 6.4 \times 10^{-3}$

$L_{ic} = 14.941$

$L_{is} = 14.944$

$L_{sc} = 14.989$

$L_{ss} = 14.986$

Limits of warning zone:

$TI - 6\sigma_0 = 31.4 \times 10^{-3}$

$a = 1.18 \times 10^{-3}$

Time between re-setting:

$\frac{3.18}{1.18} = 27$ min

9

(i) (a) χ^2 is a more rigorous test for goodness-of-fit to the normal law; however, m and σ have to be estimated in order to apply this, and here $\bar{X}_p = 450$ g, $\sigma_p = 4.97$ g, $\bar{X} = \bar{X}_p + \bar{X}_b$ The results are gaussian

(b) $\hat{\sigma} = \dfrac{\overline{W}}{dn}$; $\sigma_{n-1} = \left[\dfrac{1}{n-1}\sum_i (X_i - \overline{X})^2\right]^{1/2}$

and $\hat{\sigma} = \dfrac{1}{m}\sum_{j=1}^{j=m} \sigma_{n-1}$

$\sigma_T = (\sigma_p^2 + \sigma_b^2)^{1/2}$

(ii) (a) $U_1 = \dfrac{438 - 445}{4.97} = -1.4$

Thus 8% are outside the tolerance limits.

(b) $\overline{X} = 4489$

(iii) Control chart for the mean (product + box)

$L_{sc} = 525.6$
$L_{ss} = 522.7$
$L_{is} = 501.3$
$L_{ic} = 498.4$

Control chart for product + box

$L_{sc} = 9.6$
$L_{ss} = 7.46$

$\overline{X}_1 = 512.2$ g $\hat{\sigma}_1 = 3$ g
$\overline{X}_2 = 514.8$ g $\hat{\sigma}_2 = 4.14$ g
$\overline{X}_3 = 511.4$ g $\hat{\sigma}_3 = 4.7$ g
$\overline{X}_4 = 509$ g $\hat{\sigma}_4 = 7.92$ g

Take another sample immediately after the fourth, to check the stability of the standard deviation.

(iv) Base the calculation on 8% of boxes being outside the tolerance limits: number of boxes rejected = $5 \times 10^5 \times 8\% = 4 \times 10^4$

Cost: $4 \times 10^4 \times 445 \dfrac{15}{1000} = £267\,000$

(v) Machine A, loss 0%
Machine B, loss 1%, cost £33395.
Choose machine A.

10

(i) (a) 0.003
(b) 0.127
(c) 0.097
(d) 0.124

(ii) (a) 2.96%
(b) 74.8%

(iii) (a) 0.19
(b) 0.87

(iv) 14.3%

(v) (a) 179.4 days

(b) 0.46
(c) 150 days

11
(i) $\gamma = 0$; $\beta = 1.6$; $\eta = 225$
$\beta = 1.6 \Rightarrow A = 0.8966$
MTBF = 0.8966 225 = 201 h
(ii) $\beta > 1 \Rightarrow$ preventive maintenance

12
(i)
Graphical solution
(ii)
(iii) MTBF = 120 days

13
(i) $R(t) = \exp[-(t/370)^{2.8}]$
(ii) $t = 330$
(iii) Rate increasing in the 'aged' period

14
(i) (a) Weibull's law with $\gamma > 0$
(b) $\eta = 2900$ h; $\beta = 1.5$; $\gamma = 1200$ h
(c) $E(t) = \text{MTBF} = 1200 + 2597 = 3797$ h
(ii) (a) $P_1 = \text{NQA}$
$\alpha = 2\%$
$\beta = 0.676$
(b) Graph as shown

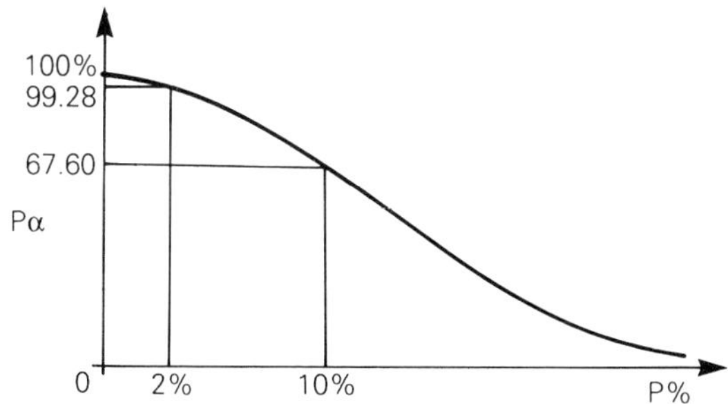

(iii) (a)

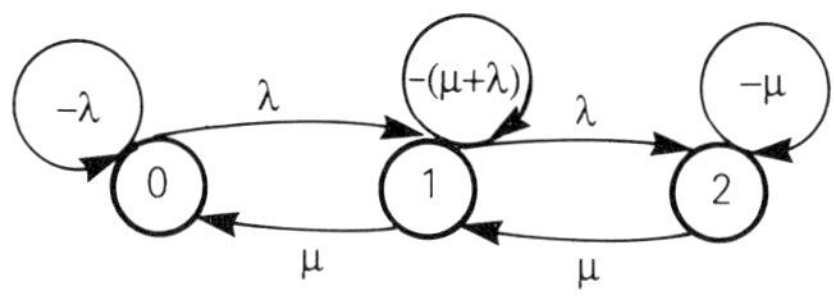

(b) $P_0'(t) = \lambda P_0(t) + \mu P_1(t)$
$P_1'(t) = \lambda P_0(t) - (\mu + \lambda) P_1(t) + \mu P_2(t)$
$P_2'(t) = \lambda P_1(t) - \mu P_2(t)$

(c) $$\begin{pmatrix} (s+\lambda) & (-\mu) & 0 \\ -\lambda & (s+\lambda+\mu) & (-\mu) \\ 0 & (-\lambda) & (s+\mu) \end{pmatrix} \begin{pmatrix} 0 \\ 1 \\ 2 \end{pmatrix} = \begin{pmatrix} 1 \\ 0 \\ 0 \end{pmatrix}$$

[s is used here for the parameter of the Laplace transform; in the text and the table, p]

(d) $$A(t) = \frac{\mu^2 + \lambda\mu}{\lambda^2 + \mu^2 + \lambda\mu} + \frac{\lambda^2}{\lambda^2 + \mu^2 + \lambda\mu}\left[\frac{C_1 \exp(+r_2 t) - C_2 \exp(r_1 t)}{C_1 + C_2}\right]$$

(e) $$\frac{\mu^2 + \lambda\mu}{\lambda^2 + \mu^2 + \lambda\mu}$$

(iv) (a) 0.9990
(b) $\lambda L > 10^{-7}$
(c) Active redundancy

15

(i) $\gamma = 50$; $\beta = 1.5$; $\eta = 52$
MTBF 95 h
(ii) $t_{50\%} = 88$ h
(iii)

t	60	80	100	120	130
$\lambda(t)$	0.012	0.023	0.032	0.039	0.042

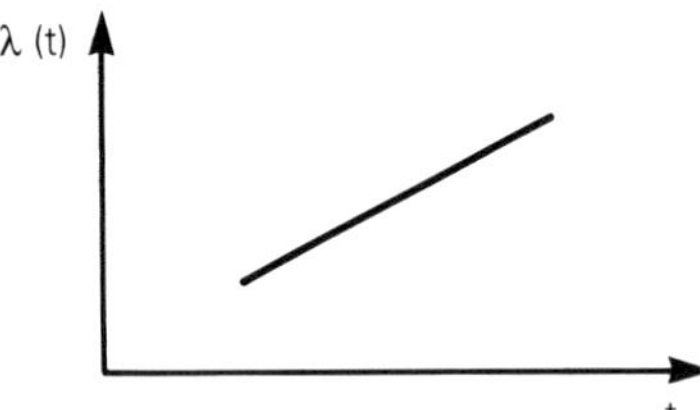

16

Time interval $t_i - t_{i+1}$	No. of failures $t_i - t_{i+1}$	No. of survivors to t_i	$\hat{R}(t_i)$	$\hat{F}(t_i)$	$\hat{\lambda}(t_i)$
0–500	3	55	1	0.054	1.09×10^{-4}
500–1000	8	52	0.94	0.14	3.07×10^{-4}
1000–1500	10	44	0.8	0.18	4.5×10^{-4}

Time interval $t_i - t_{i+1}$	No. of failures $t_i - t_{i+1}$	No. of survivors to t_i	$\hat{R}(t_i)$	$\hat{F}(t_i)$	$\hat{\lambda}(t_i)$
1500–2000	12	34	0.618	0.218	7.05×10^{-4}
2000–2500	7	22	0.4	0.127	6.36×10^{-4}
2500–3000	8	15	0.27	0.145	10.4×10^{-4}
3000–3500	7	7	0.127	0.127	20×10^{-4}

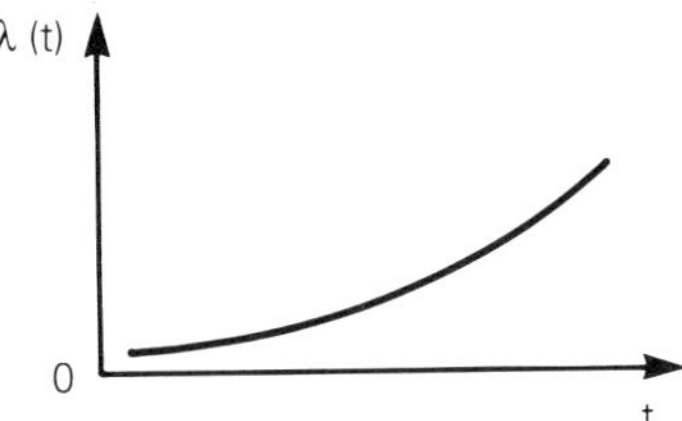

Increasing
Adopt preventive maintenance
'Ageing' period

17

(i) (a) $\gamma = 0$; $\eta = 13.8 \times 10^6$ cycles; $\beta = 2.3$

(b) Test χ^2

Number of classes ≈ 8; $E = 0.47$

$\chi^2_{8-3-1;0.05} = 9.49$: accept the model

18

$\hat{\alpha} = 15.5$

$\hat{\beta} = 19.3$

$\widehat{\alpha\beta} = 30.3$

with $Y_{..} = 140.5$

19

(i) $R(s)$ with redundancy $= 0.88$

$R(s)$ without redundancy $= 0.66$

(ii) $R(s) = 0.96$

Appendices (Tables)

1. GAUSSIAN (NORMAL) DISTRIBUTION

The table gives the cumulative distribution function $F(u) = \int_{-\infty}^{u} f(v)\,\mathrm{d}v$ where $f(v)$ is the Gauss or normal law (p. 125).

If x is distributed normally with mean μ and standard deviation σ then u is the reduced normal variable, $u = (x - \mu)/\sigma$. $F(u)$ is the probability of finding a value less than or equal to u.

The table is for positive values of u only; values for negative u are found from

$$F(-u) = 1 - F(u)$$

eg $F(-0.94) = 1 - F(0.94) = 1 - 0.82639 = 0.17361$

The second table (p. 156) gives $1 - F(u)$ for large values of u.

u	0.00	0.01	0.02	0.03	0.04	0.05	0.06	0.07	0.08	0.09
0.0	0.500 00	0.503 99	0.507 98	0.511 97	0.511 95	0.519 94	0.523 92	0.527 90	0.531 88	0.535 86
0.1	0.539 83	0.543 80	0.547 76	0.551 72	0.555 67	0.559 62	0.563 56	0.567 50	0.571 42	0.575 35
0.2	0.579 26	0.583 17	0.587 06	0.590 95	0.594 84	0.598 71	0.602 57	0.606 42	0.610 26	0.614 09
0.3	0.617 91	0.621 72	0.625 52	0.629 30	0.633 07	0.636 83	0.640 58	0.644 31	0.648 03	0.651 73
0.4	0.655 42	0.619 10	0.662 76	0.666 40	0.670 03	0.673 65	0.677 24	0.680 82	0.684 39	0.687 93
0.5	0.691 46	0.694 97	0.698 47	0.701 94	0.705 40	0.708 84	0.712 26	0.715 66	0.719 04	0.722 40
0.6	0.725 75	0.729 07	0.732 37	0.735 65	0.738 91	0.742 15	0.745 37	0.748 57	0.751 75	0.754 90
0.7	0.758 04	0.761 15	0.764 24	0.767 31	0.770 35	0.773 37	0.776 37	0.779 36	0.782 30	0.785 24
0.8	0.788 14	0.791 03	0.793 89	0.796 73	0.799 55	0.802 34	0.805 11	0.807 85	0.810 57	0.813 27
0.9	0.815 94	0.818 59	0.821 21	0.823 81	0.826 39	0.828 94	0.831 47	0.833 98	0.836 46	0.838 91
1.0	0.841 34	0.843 75	0.846 14	0.848 50	0.850 83	0.853 14	0.855 43	0.857 69	0.859 93	0.862 14
1.1	0.864 33	0.866 50	0.868 64	0.870 76	0.872 86	0.874 93	0.876 98	0.879 00	0.881 00	0.882 98
1.2	0.884 93	0.886 86	0.888 77	0.890 65	0.892 51	0.894 35	0.896 17	0.897 96	0.899 73	0.901 47
1.3	0.903 20	0.904 90	0.906 58	0.908 24	0.909 88	0.911 49	0.913 09	0.914 66	0.916 21	0.917 74
1.4	0.919 24	0.920 73	0.922 20	0.923 64	0.925 07	0.926 47	0.927 86	0.929 22	0.930 56	0.931 89
1.5	0.933 19	0.934 48	0.935 74	0.936 99	0.938 22	0.939 43	0.940 62	0.941 79	0.942 95	0.944 08
1.6	0.945 20	0.946 30	0.947 38	0.948 45	0.949 50	0.950 53	0.951 54	0.952 54	0.953 52	0.954 49
1.7	0.955 43	0.956 37	0.957 28	0.958 19	0.959 07	0.959 94	0.960 80	0.961 64	0.962 46	0.963 27
1.8	0.964 07	0.964 85	0.965 62	0.966 38	0.967 12	0.967 84	0.968 56	0.969 26	0.969 95	0.970 62
1.9	0.971 28	0.971 93	0.972 57	0.973 20	0.973 81	0.974 41	0.975 00	0.975 58	0.976 15	0.976 70

u	0.00	0.01	0.02	0.03	0.04	0.05	0.06	0.07	0.08	0.09
2.0	0.977 25	0.977 73	0.978 31	0.978 82	0.979 32	0.979 82	0.980 30	0.980 77	0.981 24	0.981 69
2.1	0.982 14	0.982 57	0.983 00	0.983 41	0.983 82	0.984 22	0.984 61	0.985 00	0.985 37	0.985 74
2.2	0.986 10	0.986 45	0.986 79	0.987 13	0.987 45	0.987 78	0.988 09	0.988 40	0.988 70	0.988 99
2.3	0.989 28	0.989 56	0.989 83	0.990 10	0.990 36	0.990 61	0.990 86	0.991 11	0.991 34	0.991 58
2.4	0.991 80	0.992 02	0.992 24	0.992 45	0.992 66	0.992 86	0.993 05	0.993 24	0.993 43	0.993 61
2.5	0.993 79	0.993 96	0.994 13	0.994 30	0.994 46	0.994 61	0.994 77	0.994 92	0.995 06	0.995 20
2.6	0.995 34	0.995 47	0.995 60	0.995 73	0.995 85	0.995 98	0.996 09	0.996 21	0.996 32	0.996 43
2.7	0.996 53	0.996 64	0.996 74	0.996 83	0.996 93	0.997 02	0.997 11	0.997 20	0.997 28	0.997 36
2.8	0.997 44	0.997 52	0.997 60	0.997 67	0.997 74	0.997 81	0.997 88	0.997 95	0.998 01	0.998 07
2.9	0.998 13	0.998 19	0.998 25	0.998 31	0.998 36	0.998 41	0.998 46	0.998 51	0.998 56	0.998 61

u	0.0	0.1	0.2	0.3	0.4	0.5	0.6	0.7	0.8	0.9
3.	$135\ 10^{-5}$	$968\ 10^{-6}$	$687\ 10^{-4}$	$683\ 10^{-6}$	$337\ 10^{-6}$	$223\ 10^{-6}$	$159\ 10^{-6}$	$108\ 10^{-6}$	$723\ 10^{-7}$	$481\ 10^{-7}$
4.	$317\ 10^{-7}$	$207\ 10^{-7}$	$133\ 10^{-7}$	$85\ 10^{-7}$	$54\ 10^{-7}$	$34\ 10^{-7}$	$21\ 10^{-7}$	$13\ 10^{-7}$	$79\ 10^{-8}$	$48\ 10^{-8}$
4.	$29\ 10^{-8}$	$17\ 10^{-8}$	$10\ 10^{-8}$	$58\ 10^{-9}$	$33\ 10^{-9}$	$19\ 10^{-9}$	$11\ 10^{-9}$	$60\ 10^{-10}$	$33\ 10^{-10}$	$18\ 10^{-10}$

2. STUDENT *t* DISTRIBUTION

$$F(t) = \frac{1}{\sqrt{2\pi}} \cdot \frac{\Gamma\left(\frac{\nu+1}{2}\right)}{\Gamma\left(\frac{\nu}{2}\right)} \cdot \left(1 + \frac{t^2}{\nu}\right)^{-\nu+1/2} \qquad F(t) = \int_{-\infty}^{t} F(\tau)\,\mathrm{d}\tau$$

Values (positive or negative) of t having probability α of being exceeded

t \ α	0.45	0.40	0.35	0.30	0.25	0.20	0.15	0.10	0.05	0.025	0.01	0.005	0.0005
1	0.158	0.325	0.510	0.727	1.000	1.376	1.963	3.078	6.314	12.706	31.821	63.657	636.619
2	0.142	0.289	0.445	0.617	0.816	1.061	1.386	1.886	2.920	4.303	6.965	9.925	31.598
3	0.137	0.277	0.424	0.584	0.767	0.978	1.250	1.638	2.353	3.182	4.541	5.841	12.929
4	0.134	0.271	0.414	0.569	0.741	0.941	1.190	1.533	2.132	2.776	3.747	4.604	8.610
5	0.132	0.267	0.408	0.559	0.727	0.920	1.156	1.476	2.015	2.571	3.365	4.032	6.869
6	0.131	0.265	0.404	0.553	0.718	0.906	1.134	1.440	1.943	2.447	3.143	3.707	5.959
7	0.130	0.263	0.402	0.549	0.711	0.896	1.119	1.415	1.895	2.365	2.998	3.499	5.408
8	0.130	0.262	0.399	0.546	0.706	0.889	1.108	1.397	1.860	2.306	2.896	3.355	5.041
9	0.129	0.261	0.398	0.543	0.703	0.883	1.100	1.383	1.833	2.262	2.821	3.250	4.781
10	0.129	0.260	0.397	0.542	0.700	0.879	1.093	1.372	1.812	2.228	2.764	3.169	4.587
11	0.129	0.260	0.396	0.540	0.697	0.876	1.088	1.363	1.796	2.201	2.718	3.106	4.437
12	0.128	0.259	0.395	0.539	0.695	0.873	1.083	1.356	1.782	2.179	2.681	3.055	4.318
13	0.128	0.259	0.394	0.538	0.694	0.870	1.079	1.350	1.771	2.160	2.650	3.012	4.221
14	0.128	0.258	0.393	0.537	0.692	0.868	1.076	1.345	1.761	2.145	2.624	2.977	4.140
15	0.128	0.258	0.393	0.536	0.691	0.866	1.074	1.341	1.753	2.131	2.602	2.947	4.073
16	0.128	0.258	0.392	0.535	0.690	0.865	1.071	1.337	1.746	2.120	2.583	2.921	4.015
17	0.128	0.257	0.392	0.534	0.689	0.863	1.069	1.333	1.740	2.110	2.567	2.898	3.965
18	0.127	0.257	0.392	0.534	0.688	0.862	1.067	1.330	1.734	2.101	2.552	2.878	3.922
19	0.127	0.257	0.391	0.533	0.688	0.861	1.066	1.328	1.729	2.093	2.539	2.861	3.883
20	0.127	0.257	0.391	0.533	0.687	0.860	1.064	1.325	1.725	2.086	2.528	2.845	3.850

t \ α	0.45	0.40	0.35	0.30	0.25	0.20	0.15	0.10	0.05	0.025	0.01	0.005	0.0005
21	0.127	0.257	0.391	0.532	0.686	0.859	1.063	1.323	1.721	2.080	2.518	2.831	3.819
22	0.127	0.256	0.390	0.532	0.686	0.858	1.061	1.321	1.717	2.074	2.508	2.819	3.792
23	0.127	0.256	0.390	0.532	0.685	0.858	1.060	1.319	1.714	2.069	2.500	2.807	3.767
24	0.127	0.256	0.390	0.531	0.685	0.857	1.059	1.318	1.711	2.064	2.492	2.797	3.745
25	0.127	0.256	0.390	0.531	0.684	0.856	1.058	1.316	1.708	2.060	2.485	2.787	3.725
26	0.127	0.256	0.390	0.531	0.684	0.856	1.058	1.315	1.706	2.056	2.479	2.779	3.707
27	0.127	0.256	0.389	0.531	0.684	0.855	1.057	1.314	1.703	2.052	2.473	2.771	3.690
28	0.127	0.256	0.389	0.530	0.683	0.855	1.056	1.289	1.701	2.048	2.467	2.763	3.674
29	0.127	0.256	0.389	0.530	0.683	0.854	1.055	1.282	1.699	2.045	2.462	2.756	3.659
30	0.127	0.256	0.389	0.530	0.683	0.854	1.055	1.313	1.697	2.042	2.457	2.750	3.646
40	0.126	0.255	0.388	0.529	0.681	0.851	1.050	1.311	1.684	2.021	2.423	2.704	3.551
80	0.126	0.254	0.387	0.527	0.679	0.848	1.046	1.310	1.671	2.000	2.390	2.660	3.460
120	0.126	0.254	0.386	0.526	0.677	0.845	1.041	1.303	1.658	1.980	2.358	2.617	3.373
∞	0.126	0.253	0.385	0.524	0.674	0.842	1.036	1.296	1.645	1.960	2.326	2.576	3.291

3. χ^2 DISTRIBUTION

$$\chi_\nu^2 = \frac{X_1 - m_1^2}{\sigma_1} + \frac{X_2 - m_2^2}{\sigma_2} + \ldots + \frac{X_\nu - m_\nu^2}{\sigma_\nu}$$

$$F(\chi_\nu^2) = \frac{1}{2(\nu/2)\Gamma(\nu/2)} \cdot (\chi_\nu^2)^{\nu/2-1} \exp\left(-\frac{\chi_\nu^2}{2}\right)$$

For $\nu > 30$ the quantity $(2\chi^2)^{1/2} - (2\nu - 1)^{1/2}$ can be taken to be a reduced normal variable.

ν \ α	0.995	0.990	0.975	0.950	0.900	0.80	0.70	0.50	0.30	0.20	0.10	0.05	0.025	0.010	0.005	0.001
1		0.0002	0.0010	0.0039	0.0158	0.0642	0.148	0.455	1.07	1.64	2.71	3.84	5.02	6.63	7.88	10.8
2	0.0100	0.0201	0.0506	0.103	0.211	0.446	0.713	1.39	2.41	3.22	4.61	5.99	7.38	9.21	10.6	13.8
3	0.0717	0.115	0.216	0.352	0.584	1.01	1.42	2.37	3.67	4.64	6.25	7.82	9.35	11.3	12.8	16.3
4	0.207	0.297	0.484	0.711	1.06	1.65	2.20	3.36	4.88	5.99	7.78	9.59	11.1	13.3	14.9	18.5
5	0.412	0.554	0.831	1.15	1.61	2.34	3.00	4.35	6.06	7.29	9.24	11.1	12.8	15.1	16.7	20.5
6	0.676	0.872	1.24	1.64	2.20	3.07	3.83	5.35	7.23	8.56	10.6	12.6	14.4	16.8	18.5	22.5
7	0.989	1.24	1.69	2.17	2.83	3.82	4.67	6.35	8.38	9.80	12.0	14.1	16.0	18.5	20.3	24.3
8	1.34	1.65	2.18	2.73	3.49	4.59	5.53	7.34	9.52	11.0	13.4	15.5	17.5	20.1	22.0	26.1
9	1.73	2.09	2.70	3.33	4.17	5.38	6.39	8.34	10.7	12.2	14.7	16.9	19.0	21.7	23.6	27.9
10	2.16	2.56	3.25	3.94	4.87	6.18	7.27	9.34	11.8	13.4	16.0	18.3	20.5	23.2	25.2	29.6
11	2.60	3.05	3.82	4.57	5.58	6.99	8.15	10.3	12.9	14.6	17.3	19.7	21.9	24.7	26.8	31.3
12	3.07	3.57	4.40	5.23	6.30	7.81	9.03	11.3	14.0	15.8	18.5	21.0	23.3	26.2	28.3	32.9
13	3.57	4.11	5.01	5.89	7.04	8.63	9.93	12.3	15.1	17.0	19.8	22.4	24.7	27.7	29.8	34.5
14	4.07	4.66	5.63	6.57	7.79	9.47	10.8	13.3	16.2	18.2	21.1	23.7	26.1	29.1	31.3	36.1
15	4.60	5.23	6.26	7.26	8.55	10.3	11.7	14.3	17.3	19.3	22.3	25.0	27.5	30.6	32.8	37.7
16	5.14	5.81	6.91	7.96	9.31	11.2	12.6	15.3	18.4	20.5	23.5	26.3	28.8	32.0	34.3	39.3
17	5.70	6.41	7.56	8.67	10.1	12.0	13.5	16.3	19.5	21.6	24.8	27.6	30.2	33.4	35.7	40.8
18	6.26	7.01	8.23	9.39	10.9	12.9	14.4	17.3	20.6	22.8	26.0	28.9	31.5	34.8	37.2	42.3
19	6.84	7.63	8.91	10.1	11.7	13.7	15.4	18.3	21.7	23.9	27.2	30.1	32.9	36.2	38.6	43.8
20	7.43	8.26	9.59	10.9	12.4	14.6	16.3	19.3	22.8	25.0	28.4	31.4	34.2	37.6	40.0	45.3

ν \ α	0.995	0.990	0.975	0.950	0.900	0.80	0.70	0.50	0.30	0.20	0.10	0.05	0.025	0.010	0.005	0.001
21	8.03	8.90	10.3	11.6	13.2	15.4	17.2	20.3	23.9	26.2	29.6	32.7	35.5	38.9	41.4	46.8
22	8.64	9.54	11.0	12.3	14.0	16.3	18.1	21.3	24.9	27.3	30.8	33.9	36.8	40.3	42.8	48.3
23	9.26	10.2	11.7	13.1	14.8	17.2	19.0	22.3	26.0	28.4	32.0	35.2	38.1	41.6	44.2	49.7
24	9.89	10.9	12.4	13.8	15.7	18.1	19.9	23.3	27.1	29.6	33.2	36.4	39.4	43.0	45.6	51.2
25	10.5	11.5	13.1	14.6	16.5	18.9	20.9	24.3	28.2	30.7	34.4	37.7	40.6	44.3	46.9	52.6
26	11.2	12.2	13.8	15.4	17.3	19.8	21.8	25.3	29.2	31.8	35.6	38.9	41.9	45.6	48.3	54.1
27	11.8	12.9	14.6	16.2	18.1	20.7	22.7	26.3	30.3	32.9	36.7	40.1	43.2	47.0	49.6	55.5
28	12.5	13.6	15.3	16.9	18.9	21.6	23.6	27.3	31.4	34.0	37.9	41.3	44.5	48.3	51.0	56.9
29	13.1	14.3	16.0	17.7	19.8	22.5	24.6	28.3	32.5	35.1	39.1	42.6	45.7	49.6	52.3	58.3
30	13.8	15.0	16.8	18.5	20.6	23.4	25.5	29.3	33.5	36.3	40.3	43.8	47.0	50.9	53.7	59.7

4. THE F (FISHER–SNEDECOR) DISTRIBUTION

$$F = \frac{\chi_1^2/\nu_1}{\chi_2^2/\nu_2}$$

χ_1^2, χ_2^2 have ν_1, ν_2 degrees of freedom respectively. The table gives values of F having probability α of being exceeded.

	$\nu_1 = 1$		$\nu_1 = 2$		$\nu_1 = 3$		$\nu_1 = 4$		$\nu_1 = 5$	
ν_2	α 0.05	0.01	0.05	0.01	0.05	0.01	0.05	0.01	0.05	0.01
1	161.4	4052	199.5	4999	215.7	5403	224.6	5625	230.2	5764
2	18.51	98.49	19.00	99.00	19.16	99.17	19.25	99.25	19.30	99.30
3	10.13	34.12	9.55	30.81	9.28	29.46	9.12	28.71	9.01	28.24
4	7.71	21.20	6.94	18.00	6.59	16.69	6.39	15.98	6.26	15.52
5	6.61	16.26	5.79	13.27	5.41	12.60	5.19	11.39	5.05	10.97
6	5.99	13.74	5.14	10.91	4.76	9.78	4.53	9.15	4.39	8.75
7	5.59	12.25	4.74	9.55	4.35	8.45	4.12	7.85	3.97	7.45
8	5.32	11.26	4.46	8.65	4.07	7.59	3.84	7.01	3.69	6.63
9	5.12	10.56	4.26	8.02	3.86	6.99	3.63	6.42	3.48	6.06
10	4.96	10.04	4.10	7.56	3.71	6.55	3.48	5.99	3.33	5.64
11	4.84	9.65	3.98	7.20	3.59	6.22	3.36	5.67	3.20	5.32
12	4.75	9.33	3.88	6.93	3.49	5.95	3.26	5.41	3.11	5.06
13	4.67	9.07	3.80	6.70	3.41	5.74	3.18	5.20	3.02	4.86
14	4.60	8.86	3.74	6.51	3.34	5.56	3.11	5.03	2.96	4.69
15	4.54	8.68	3.68	6.36	3.29	5.42	3.06	4.89	2.90	4.56
16	4.49	8.53	3.63	6.23	3.24	5.29	3.01	4.77	2.85	4.44
17	4.45	8.40	3.59	6.11	3.20	5.18	2.96	4.67	2.81	4.34
18	4.41	8.28	3.55	6.01	3.16	5.09	2.93	4.58	2.77	4.25
19	4.38	8.18	3.52	5.93	3.13	5.01	2.90	4.50	2.74	4.17
20	4.35	8.10	3.49	5.85	3.10	4.94	2.87	4.43	2.71	4.10

	$\nu_1 = 1$		$\nu_1 = 2$		$\nu_1 = 3$		$\nu_1 = 4$		$\nu_1 = 5$	
ν_2	α 0.05	0.01	0.05	0.01	0.05	0.01	0.05	0.01	0.05	0.01
21	4.32	8.02	3.47	5.78	3.07	4.87	2.84	4.37	2.68	4.04
22	4.30	7.94	3.44	5.72	3.05	4.82	2.82	4.31	2.66	3.99
23	4.28	7.88	3.42	5.66	3.03	4.76	2.80	4.26	2.64	3.94
24	4.26	7.82	3.40	5.61	3.01	4.72	2.78	4.22	2.62	3.90
25	4.24	7.77	3.38	5.57	2.99	4.68	2.76	4.18	2.60	3.86
26	4.22	7.72	3.37	5.53	2.98	4.64	2.74	4.14	2.59	3.82
27	4.21	7.68	3.35	5.49	2.96	4.60	2.73	4.11	2.57	3.78
28	4.20	7.64	3.34	5.45	2.95	4.57	2.71	4.07	2.56	3.75
29	4.18	7.60	3.33	5.42	2.93	4.54	2.70	4.04	2.54	3.73
30	4.17	7.56	3.32	5.39	2.92	4.51	2.69	4.02	2.53	3.70
40	4.08	7.31	3.23	5.18	2.84	4.31	2.61	3.83	2.45	3.51
60	4.00	7.08	3.15	4.98	2.76	4.13	2.52	3.65	2.37	3.34
120	3.92	6.85	3.07	4.79	2.68	3.95	2.45	3.48	2.29	3.17
∞	3.84	6.64	2.99	4.60	2.60	3.78	2.37	3.32	2.21	3.02

	$\nu_1 = 6$		$\nu_1 = 8$		$\nu_1 = 12$		$\nu_1 = 24$		$\nu_1 = \infty$	
ν_2	α 0.05	0.01	0.05	0.01	0.05	0.01	0.05	0.01	0.05	0.01
1	234.0	5 859	238.9	5 981	243.9	6 106	249.0	6 234	254.3	6 366
2	19.33	99.33	19.37	99.36	19.41	99.42	19.45	99.46	19.50	99.50
3	8.94	27.91	8.84	27.49	8.74	27.05	8.64	26.60	8.53	26.12
4	6.16	15.21	6.04	14.80	5.91	14.37	5.77	13.93	5.63	13.46
5	4.95	10.67	4.82	10.27	4.68	9.89	4.53	9.47	4.36	9.02
6	4.28	8.47	4.15	8.10	4.00	7.72	3.84	7.31	3.67	6.88
7	3.87	7.19	3.73	6.84	3.57	6.47	3.41	6.07	3.23	5.65
8	3.58	6.37	3.44	6.03	3.28	5.67	3.12	5.28	2.93	4.86
9	3.37	5.80	3.23	5.47	3.07	5.11	2.90	4.73	2.71	4.31
10	3.22	5.39	3.07	5.06	2.91	4.71	2.74	4.33	2.54	3.91
11	3.09	5.07	2.95	4.74	2.79	4.40	2.61	4.02	2.40	3.60
12	3.00	4.82	2.85	4.50	2.69	4.16	2.50	3.78	2.30	3.36
13	2.92	4.62	2.77	4.30	2.60	3.96	2.42	3.59	2.21	3.16
14	2.85	4.46	2.70	4.14	2.53	3.80	2.35	3.43	2.13	3.00
15	2.79	4.32	2.64	4.00	2.48	3.67	2.29	3.29	2.07	2.87
16	2.74	4.20	2.59	3.89	2.42	3.55	2.24	3.18	2.01	2.75
17	2.70	4.10	2.55	3.79	2.38	3.45	2.19	3.08	1.96	2.65
18	2.66	4.01	2.51	3.71	2.34	3.37	2.15	3.00	1.92	2.57
19	2.63	3.94	2.48	3.63	2.31	3.30	2.11	2.92	1.88	2.49
20	2.60	3.87	2.45	3.56	2.28	3.23	2.08	2.86	1.84	2.42

	$v_1 = 6$		$v_1 = 8$		$v_1 = 12$		$v_1 = 24$		$v_1 = \infty$	
v_2	α 0.05	0.01	0.05	0.01	0.05	0.01	0.05	0.01	0.05	0.01
21	2.57	3.81	2.42	3.51	2.25	3.17	2.05	2.80	1.81	2.36
22	2.55	3.76	2.40	3.45	2.23	3.12	2.03	2.75	1.78	2.31
23	2.53	3.71	2.38	3.41	2.20	3.07	2.00	2.70	1.76	2.26
24	2.51	3.67	2.36	3.36	2.18	3.03	1.98	2.66	1.73	2.21
25	2.49	3.63	2.34	3.32	2.16	2.99	1.96	2.62	1.71	2.17
26	2.47	3.59	2.32	3.29	2.15	2.96	1.95	2.58	1.69	2.13
27	2.46	3.56	2.30	3.26	2.13	2.93	1.93	2.55	1.67	2.10
28	2.44	3.53	2.29	3.23	2.12	2.90	1.91	2.52	1.65	2.06
29	2.43	3.50	2.28	3.20	2.10	2.87	1.90	2.49	1.64	2.03
30	2.42	3.47	2.27	3.17	2.09	2.84	1.89	2.47	1.62	2.01
40	2.34	3.29	2.18	2.99	2.00	2.66	1.79	2.29	1.51	1.80
60	2.25	3.12	2.10	2.82	1.92	2.50	1.70	2.12	1.39	1.60
120	2.17	2.96	2.01	2.66	1.83	2.34	1.61	1.95	1.25	1.38
∞	2.09	2.80	1.94	2.51	1.75	2.18	1.52	1.79	1.00	1.00

5. MEAN TIME BETWEEN FAILURES FOR A SYSTEM FOLLOWING THE WEIBULL LAW

The table gives the values of A, B where the mean and standard deviation of the MTBF are found from

$$\text{mean} = A\eta + \gamma \qquad \text{standard deviation} = B\eta$$

β	A	B
0.20	120	1901
0.25	24	199
0.30	9.2605	50.08
0.35	5.0791	19.98
0.40	3.3234	10.44
0.45	2.4786	6.46
0.50	2.0000	4.47
0.55	1.7024	3.35
0.60	1.5046	2.65
0.65	1.3663	2.18
0.70	1.2638	1.85
0.75	1.1906	1.61
0.80	1.1330	1.43
0.85	1.0880	1.29
0.90	1.0522	1.17
0.95	1.0234	1.08
1.00	1.0000	1.00
1.05	0.9803	0.934
1.10	0.9649	0.878
1.15	0.9517	0.830
1.20	0.9407	0.787
1.25	0.9314	0.750
1.30	0.9236	0.716
1.35	0.9170	0.687
1.40	0.9114	0.660
1.45	0.9067	0.635
1.50	0.9027	0.613
1.55	0.8994	0.593
1.60	0.8966	0.574
1.65	0.8942	0.556
1.70	0.8922	0.540
1.75	0.8906	0.525

β	A	B
1.80	0.8893	0.511
1.85	0.8882	0.498
1.90	0.8874	0.486
1.95	0.8867	0.474
2.0	0.8862	0.463
2.1	0.8857	0.443
2.2	0.8856	0.425
2.3	0.8859	0.409
2.4	0.8865	0.393
2.5	0.8873	0.380
2.6	0.8882	0.367
2.7	0.8893	0.355
2.8	0.8905	0.344
2.9	0.8917	0.334
3	0.8930	0.325
3.1	0.8943	0.316
3.2	0.8957	0.307
3.3	0.8970	0.299
3.4	0.8984	0.292
3.5	0.8997	0.285
3.6	0.9011	0.278
3.7	0.9025	0.272
3.8	0.9038	0.266
3.9	0.9051	0.260
4	0.9064	0.254
4.1	0.9077	0.249
4.2	0.9089	0.244
4.3	0.9102	0.239
4.4	0.9144	0.235
4.5	0.9126	0.230
4.6	0.9137	0.226
4.7	0.9149	0.222
4.8	0.9160	0.218
4.9	0.9171	0.214
5	0.9182	0.210
5.1	0.9192	0.207

β	A	B
5.2	0.9202	0.203
5.3	0.9213	0.200
5.4	0.9222	0.197
5.5	0.9232	0.194
5.6	0.9241	0.191
5.7	0.9251	0.188
5.8	0.9260	0.185
5.9	0.9269	0.183
6	0.9277	0.180
6.1	0.9286	0.177
6.2	0.9294	0.175
6.3	0.9302	0.172
6.4	0.9310	0.170
6.5	0.9316	0.168
6.6	0.9325	0.166
6.7	0.9333	0.163
6.8	0.9340	0.161
6.9	0.9347	0.160

6. MEDIAN RANKS (JOHNSON'S TABLE)

Rank order	Sample size 1	2	3	4	5	6	7	8	9	10
1	50.000	29.289	20.630	15.910	12.945	10.910	9.428	8.300	7.412	6.697
2		70.711	50.000	38.573	31.381	26.445	22.849	20.113	17.962	16.226
3			79.370	61.427	50.000	42.141	36.412	32.052	28.624	25.857
4				84.090	68.619	57.859	50.000	44.015	39.308	35.510
5					87.055	73.555	63.588	55.984	50.000	45.169
6						89.090	77.151	67.948	60.691	54.831
7							90.572	79.887	71.376	64.490
8								91.700	82.038	74.142
9									92.587	83.774
10										93.303

Rank order	Sample size									
	11	12	13	14	15	16	17	18	19	20
1	6.107	5.613	5.192	4.830	4.516	4.240	3.995	3.778	3.582	3.406
2	14.796	13.598	12.579	11.702	10.940	10.270	9.678	9.151	8.677	8.251
3	23.578	21.669	20.045	18.647	17.432	16.365	15.422	14.581	13.827	13.147
4	32.380	29.758	27.528	25.608	23.939	22.474	21.178	20.024	18.988	18.055
5	41.189	37.853	35.016	32.575	30.452	28.589	26.940	25.471	24.154	22.967
6	50.000	45.951	42.508	39.544	36.967	34.705	32.704	30.921	29.322	27.880
7	58.811	54.049	50.000	46.515	43.483	40.823	38.469	36.371	34.491	32.795
8	67.620	62.147	57.492	53.485	50.000	46.941	44.234	41.823	39.660	37.710
9	76.421	70.242	64.984	60.456	56.517	53.059	50.000	47.274	44.830	42.626
10	85.204	78.331	72.472	67.425	63.033	59.177	55.766	52.726	50.000	47.542
11	93.893	86.402	79.955	74.392	69.548	65.295	61.531	58.177	55.170	52.458
12		94.387	87.421	81.353	76.061	71.411	67.296	63.629	60.340	57.374
13			94.808	88.298	82.568	77.525	73.060	69.079	65.509	62.289
14				95.169	89.060	83.635	78.821	74.629	70.678	67.205
15					95.484	88.730	84.578	79.976	75.846	72.119
16						95.760	90.322	85.419	81.011	77.033
17							96.005	90.849	86.173	81.945
18								96.222	91.322	86.853
19									96.418	91.749
20										96.594

Rank order	Sample size									
	21	22	23	24	25	26	27	28	29	30
1	3.247	3.101	2.969	2.847	2.734	2.631	2.534	2.445	2.362	2.284
2	7.864	7.512	7.191	6.895	6.623	6.372	6.139	5.922	5.720	5.532
3	12.531	11.970	11.458	10.987	10.553	10.153	9.781	9.436	9.114	8.814
4	17.209	16.439	15.734	15.088	14.492	13.942	13.432	12.958	12.517	12.104
5	21.890	20.911	20.015	19.192	18.435	17.735	17.086	16.483	15.922	15.397
6	26.574	25.384	24.297	23.299	22.379	21.529	20.742	20.010	19.328	18.691
7	31.258	29.859	28.580	27.406	26.324	25.325	24.398	23.537	22.735	21.986
8	35.943	34.334	32.863	31.513	30.269	29.120	28.055	27.065	26.143	25.281
9	40.629	38.810	37.147	35.621	34.215	32.916	31.712	30.593	29.550	28.576
10	45.314	43.286	41.431	39.729	38.161	36.712	35.370	34.121	32.958	31.872
11	50.000	47.762	45.716	43.837	42.107	40.509	39.027	37.650	36.367	35.168
12	54.686	52.238	50.000	47.946	46.054	44.305	42.685	41.178	39.775	38.464
13	59.371	56.714	54.284	52.054	50.000	48.102	46.342	44.707	43.183	41.760
14	64.057	61.190	58.568	56.162	53.946	51.898	50.000	48.236	46.592	45.056
15	68.742	65.665	62.853	60.271	57.892	55.695	53.658	51.764	50.000	48.352
16	73.426	70.141	67.137	64.379	61.839	59.491	57.315	55.293	53.408	51.648
17	78.109	74.616	71.420	68.487	65.785	63.287	60.973	58.821	56.817	54.944
18	82.791	79.089	75.703	72.594	69.730	67.084	64.630	62.350	60.225	58.240
19	87.469	83.561	79.985	76.701	73.676	70.880	68.288	65.878	63.633	61.536
20	92.136	88.030	84.266	80.808	77.621	74.675	71.945	69.407	67.041	64.832

Rank order	Sample size									
	21	22	23	24	25	26	27	28	29	30
21	96.753	92.488	88.542	84.912	81.565	78.471	75.602	72.935	70.450	68.128
22		96.898	92.809	89.013	85.507	82.265	79.258	76.463	73.857	71.424
23			97.031	93.105	89.447	86.058	82.914	79.990	77.265	74.719
24				97.153	93.377	89.847	86.568	83.517	80.672	78.014
25					97.265	93.628	90.219	87.042	84.078	81.309
26						97.369	93.861	90.564	87.483	84.603
27							97.465	94.078	90.885	87.896
28								97.555	94.280	91.186
29									97.638	94.468
30										97.716

Rank order	Sample size 31	32	33	34	35	36	37	38	39	40
1	2.211	2.143	2.078	2.018	1.961	1.907	1.856	1.807	1.762	1.718
2	5.355	5.190	5.034	4.887	4.749	4.618	4.496	4.377	4.266	4.160
3	8.533	8.269	8.021	7.787	7.567	7.359	7.162	6.975	6.798	6.629
4	11.718	11.355	11.015	10.694	10.391	10.105	9.835	9.578	9.335	9.103
5	14.905	14.445	14.011	13.603	13.218	12.855	12.510	12.184	11.874	11.580
6	18.094	17.535	17.009	16.514	16.046	15.605	15.187	14.791	14.415	14.057
7	21.284	20.625	20.007	19.425	18.875	18.355	17.864	17.398	16.956	16.535
8	24.474	23.717	23.006	22.336	21.704	21.107	20.541	20.005	19.497	19.013
9	27.664	26.809	26.005	25.246	24.533	23.858	23.219	22.613	22.038	21.492
10	30.855	29.901	29.004	28.159	27.362	26.609	25.897	25.221	24.580	23.971
11	34.046	32.993	32.003	31.071	30.192	29.361	28.575	27.829	27.122	26.449
12	37.236	36.085	35.003	33.983	33.022	32.113	31.253	30.437	29.664	28.928
13	40.427	39.177	38.002	36.895	35.851	34.865	33.931	33.046	32.206	31.407
14	43.618	42.269	41.001	39.807	38.681	37.616	36.609	35.654	34.748	33.886
15	46.809	46.809	44.004	42.720	41.511	40.368	39.287	38.262	37.290	36.365
16	50.000	48.454	47.000	45.632	44.340	43.120	41.965	40.871	39.832	38.844
17	53.191	51.546	50.000	48.544	47.170	45.872	44.644	43.479	42.374	41.323
18	56.382	54.638	52.999	51.456	50.000	48.624	47.322	46.087	44.916	43.802
19	59.573	57.731	55.999	54.368	52.830	51.376	50.000	48.696	47.458	46.281
20	62.763	60.823	58.998	57.280	55.660	54.128	52.678	51.304	50.000	48.760

Rank order	Sample size									
	31	32	33	34	35	36	37	38	39	40
21	65.954	63.915	61.998	60.193	58.489	56.880	55.356	53.913	52.542	51.239
22	69.145	67.007	64.997	63.105	61.319	59.632	58.035	56.521	55.084	53.719
23	72.335	70.099	67.997	66.017	64.149	62.383	60.713	59.129	57.626	56.198
24	75.526	73.191	70.996	68.929	66.978	65.135	63.391	61.738	60.168	58.677
25	78.716	76.283	73.995	71.841	69.808	67.887	66.069	64.346	62.710	61.156
26	81.906	79.374	76.994	74.752	72.637	70.639	68.747	66.954	65.252	63.635
27	85.094	82.465	79.993	77.664	75.467	73.391	71.425	69.562	67.794	66.114
28	88.282	85.555	82.991	80.575	78.296	76.142	74.103	72.171	70.336	68.598
29	91.467	88.644	85.989	83.486	81.125	78.893	76.781	74.779	72.878	71.072
30	94.645	91.731	88.985	86.397	83.954	81.645	79.459	77.387	75.420	73.550
31	97.789	94.810	91.979	89.306	86.782	84.395	82.136	79.994	77.962	76.029
32		97.857	94.966	92.213	89.608	87.145	84.813	82.602	80.503	78.508
33			97.921	95.113	92.433	89.894	87.490	85.209	83.044	80.986
34				97.982	92.251	92.641	90.165	87.816	85.585	83.465
35					98.039	95.382	92.838	90.422	88.126	85.943
36						98.093	95.505	93.025	90.665	88.420
37							98.144	95.622	93.202	90.897
38								98.192	95.734	93.371
39									98.238	95.839
40										98.282

Rank order	Sample size									
	41	42	43	44	45	46	47	48	49	50
1	1.676	1.637	1.599	1.563	1.528	1.495	1.464	1.434	1.405	1.377
2	4.060	3.964	3.872	3.785	3.702	3.622	3.545	3.472	3.402	3.334
3	6.469	6.316	6.170	6.031	5.898	5.771	5.649	5.532	5.420	5.312
4	8.883	8.673	8.473	8.282	8.099	7.925	7.757	7.597	7.443	7.295
5	11.300	11.033	10.778	10.535	10.303	10.080	9.867	9.663	9.467	9.279
6	13.717	13.393	13.084	12.789	12.507	12.237	11.979	11.731	11.493	11.265
7	16.135	15.754	15.391	15.043	14.712	14.394	14.090	13.799	13.519	13.250
8	18.554	18.115	17.697	17.298	16.917	16.551	16.202	15.867	15.545	15.236
9	20.972	20.477	20.004	19.554	19.122	18.709	18.314	17.935	17.571	17.222
10	23.391	22.838	22.311	21.808	21.237	20.867	20.426	20.003	19.598	19.209
11	25.810	25.200	24.618	24.063	23.532	23.025	22.538	22.072	21.625	21.195
12	28.228	27.562	26.926	26.318	25.738	25.182	24.650	24.140	23.651	23.181
13	30.647	29.924	29.233	28.574	27.943	27.340	26.763	26.209	25.678	25.168
14	33.066	32.285	31.540	30.829	30.149	29.498	28.875	28.278	27.705	27.154
15	35.485	34.647	33.848	33.084	32.355	31.656	30.988	30.347	29.731	29.141
16	37.905	37.009	36.155	35.340	34.560	33.814	33.100	32.415	31.758	31.127
17	40.324	39.371	38.463	37.595	36.766	35.972	35.212	34.484	33.785	33.114
18	42.743	41.733	40.770	39.851	38.972	38.130	37.325	36.553	35.812	35.100
19	45.162	44.095	43.078	42.106	41.177	39.437	39.437	38.622	37.839	37.087
20	47.581	46.457	45.385	44.361	43.383	42.447	41.550	40.690	39.866	39.074
21	50.000	48.819	47.692	46.617	45.589	44.605	43.662	42.759	41.892	41.060
22	52.419	51.181	50.000	48.872	47.794	46.763	45.775	44.828	43.919	43.047
23	54.838	53.543	52.307	51.128	50.000	48.921	47.887	46.897	45.946	45.033
24	57.257	55.905	54.615	53.383	52.206	51.079	50.000	48.966	47.972	47.020
25	59.676	58.267	56.922	55.639	54.411	53.237	52.112	51.034	50.000	49.007

Rank order	Sample size									
	41	42	43	44	45	46	47	48	49	50
26	62.095	60.629	59.230	57.894	56.617	55.395	54.225	53.103	52.027	50.993
27	64.514	62.991	61.537	60.149	58.823	57.553	56.337	55.172	54.054	52.980
28	66.933	65.353	63.845	62.405	61.028	59.711	58.450	57.241	56.081	54.966
29	69.352	67.714	66.152	64.660	63.234	61.869	60.562	59.310	58.107	56.953
30	71.771	70.076	68.459	66.916	65.440	64.027	62.675	61.378	60.134	58.940
31	74.190	72.438	70.767	69.171	67.645	66.186	64.787	63.447	62.161	60.926
32	76.609	74.800	73.074	71.426	69.851	68.334	66.900	65.516	64.188	62.913
33	79.028	77.162	75.381	73.681	72.056	70.502	69.012	67.585	66.215	64.899
34	81.446	79.523	77.689	75.937	74.262	72.660	71.125	69.653	68.242	66.886
35	83.865	81.885	79.996	78.192	76.467	74.817	73.237	71.722	70.268	68.873
36	86.283	84.246	82.303	80.447	78.673	76.975	75.349	73.791	72.295	70.859
37	88.700	86.807	84.609	82.702	80.878	79.133	77.462	75.859	74.322	72.846
38	91.117	88.967	86.916	84.956	83.083	81.291	79.574	77.928	76.349	74.832
39	93.531	91.327	89.222	87.211	85.288	83.448	81.686	79.997	78.375	76.819
40	95.940	93.684	91.527	89.465	87.493	85.606	83.798	82.065	80.402	78.805
41	98.324	96.036	93.830	91.718	89.697	87.763	85.910	84.133	82.428	80.791
42		98.363	96.127	93.969	91.900	89.200	88.021	86.201	84.455	82.778
43			98.401	96.215	94.102	92.075	90.132	88.269	86.481	84.764
44				98.437	96.298	94.229	92.243	90.337	88.507	86.750
45					98.471	96.378	94.351	92.403	90.532	88.735
46						98.504	96.455	94.468	92.557	90.721
47							98.536	96.528	94.580	92.705
48								98.566	96.598	94.688
49									98.595	96.666
50										98.623

7. LAPLACE TRANSFORMS

$X(t)$	$L(p)$
Dirac impulse (δ function)	1
Unit step function	$\frac{1}{P}$
at	$\frac{a}{p^2}$
t^n	$\frac{n!}{p^{n+1}}$
$e^{-\alpha t}$	$\frac{1}{P + \alpha}$
$te^{-\alpha t}$	$\frac{1}{(P + \alpha)^2}$
$\sin \omega t$	$\frac{\omega}{\omega/p^2 + \omega^2}$
$\cos \omega t$	$\frac{p}{P^2 + \omega^2}$
$e^{-\alpha t} \sin \omega t$	$\frac{\omega}{(P + \alpha)^2 + \omega^2}$
$e^{-\alpha t} \cos \omega t$	$\frac{P + \alpha}{(P + \alpha)^2 + \omega^2}$

8. RANDOM NUMBERS

22719	92549	10907	35994	63461	83659	24494	53825	97047	76069
17618	88357	52487	79816	74600	50436	88823	19806	33960	30928
25267	35973	80231	60039	50253	63457	97444	13799	35853	03149
88594	69428	66934	27705	51262	63941	77660	66418	84755	29197
60482	33679	03078	08047	39891	34068	81957	02985	83113	36981
30753	19458	02849	30366	83892	80912	91335	41703	79401	97251
60551	24788	35764	57453	06341	10178	91896	70819	46440	98356
35612	09972	98891	92625	70599	95484	34858	13499	28966	88287
43713	18448	45922	55179	18442	31186	91047	37949	76542	79361
73998	97374	66685	06639	34590	17935	79544	15475	74765	11199
14971	68806	49122	16124	61905	22047	17229	46703	39727	16753
78976	48382	25242	97656	51686	15537	73857	35398	91783	92825
37868	82946	73732	63230	85306	56988	15570	98029	42208	00190
01666	48114	95183	02628	05355	97627	74554	91267	31240	34723
56638	70054	19427	24811	37164	71641	50515	88231	99539	75745
43973	07496	17405	08966	65989	68017	56975	94080	93689	98889
05540	72301	36504	00187	90375	22891	22205	27777	84803	39220
95141	07885	94399	41145	50210	92423	13303	09621	94153	18691
75954	68499	42308	38387	52163	64563	02843	45577	93125	25294
97905	05301	98496	20682	68082	68537	70220	78282	02396	10002

23458	57782	67537	38813	00377	93873	97813	10039	25457	28716
03954	14799	63187	46191	12805	50502	08810	19572	48024	58206
52251	06804	85959	20974	73104	15009	25486	09306	24721	04187
62361	59105	39338	59358	69193	15586	57695	89518	59788	04215
54954	90337	99346	60442	90933	58323	83183	90041	44236	90815
70773	03331	84228	01405	61494	72064	24713	39851	01431	60841
68702	08331	09823	83173	67081	87472	47980	08802	95495	78745
39599	33465	96705	41458	34670	55385	25484	71068	15155	85371
54958	34935	16858	16523	54262	63310	50348	53457	39440	80411
98124	08864	36485	78766	52802	56315	43523	06513	50899	86432
43099	88373	80091	35058	35755	47556	98602	71744	70442	92312
88667	44515	80435	17140	32588	98708	93010	98590	23656	85664
87009	95736	76930	71090	27143	95229	24799	02313	17436	20273
70581	40618	16631	54178	44737	02544	81368	08078	46740	52583
03723	25551	03816	97612	99833	06779	47619	12901	60179	23780

9. GAMMA LAW

$$f(x) = \lambda^n \exp(-\lambda x) X^{n-1}/n!$$

$$\Gamma(n) = (n - 1)! \text{ for } n \in N$$

$$\Gamma(x) = \int_0^\infty t^{x-1} \exp(-t)\, dt$$

x	$\Gamma(x)$
1.0	1.0000
1.1	0.9514
1.2	0.9182
1.3	0.8975
1.4	0.8873
1.5	0.8862
1.6	0.8935
1.7	0.9086
1.8	0.9314
1.9	0.9618
2.0	1.0000
2.1	1.0465
2.2	1.1018
2.3	1.1667
2.4	1.2422
2.5	1.3293
2.6	1.4296
2.7	1.5447
2.8	1.6765
2.9	1.8274
3.0	2.0000

Index

ABC (Pareto) analysis 101–4
Availability definition 10
  determination 38–42

Bayes theorem 119
Binomal law 74, 120

Cause-and-effect analysis 99 ff
Central limit theorem 127
Chi-square (χ^2) test 24, 59, 67, 108, Table 3
Confidence interval for mean MTBF 130, 131
Control actions 77
  BASIC program 80
  charts 61–78
    for attributes 73
      individual values 72
      mean 62
      range 69
      standard deviation 67
  levels 90
Criticality coefficient 47

Databases
  AVCO 25–32
  CNET 25–32
  NASA 25–32
  RADC 25–32
  US Navy 25–32
Drifts (in observations) 77

Experimental design 112–16
Exponential law 10, 128

Failure function
  cumulative 10
  instantaneous 10
Failure Mode Analysis (FMA) 45–55
Fisher test 110, Table 4
Function
  analysis 47
  tree 52

Goodness-of-fit tests 24, 59
  chi-square (χ^2) test 24, Table 3

Ishikawa diagram 99–101

Laplace transform 39, Table 7
log-normal law 12, 127

Monitoring 61 ff
Monte Carlo method 42
Markov chain 36
MTBF (Mean Time Between Failures) 11
MTTR (Mean Time To Repair) 38

Ownership costs 9

Pareto (ABC) analysis 101–4
Probability laws, distributions (definitions, properties)
  binomial 120
  chi-square Table 3
  exponential 128

Probability laws (*cont.*)
  Fisher Table 4
  hypergeometric 122
  log-normal 127
  normal (Gauss) 124, Table 1
  Poisson 123
  'Student' Table 2
  Weibull 129, Table 5
Probability theory 118

Quality
  assurance 4
  audit 6
  certification 2, 6
  circles 9
  control 5, 82 ff
  definitions 5
  diagnosis 8
  scales 3
  tools for achieving 7

Random numbers 43, Table 8
Rank correlation 104
  BASIC program 105
Ranks
  method of mean/median 19, Table 6
Redundancy 35
Regression
  linear 132
Reliability
  definition, function 10
  laws 10 ff
  models 11 ff
  of systems 32 ff
Risk analysis
  to customer/supplier 84

Sampling
  simple, double, multiple 82 ff
  procedures 97–8
Scatter diagram 116
Sequential testing 35, 95
Spearman (rank correlation) coefficient 105
State-change equations, graphs 37–42
'Student' test 105, Table 4

Testing
  sequential 35, 95
  truncated 35
Transition equations
  graphs 37–42

Variability
  of products 56 ff
  random 58
  systematic 59
Variance
  analysis of 106–17
  general linear model 106
  BASIC program 111
  Yates notation 113

Wald (sequential) test 95
Weibull law 12–22
  BASIC program 23, 129

Yates notation 113

IRISH MANAGEMENT
LIBRARY
INSTITUTE